VALUING CLIMATE DAMAGES
Updating Estimation of the Social Cost of Carbon Dioxide

Committee on Assessing Approaches to
Updating the Social Cost of Carbon

Board on Environmental Change and Society

Division of Behavioral and Social Sciences and Education

A Report of

The National Academies of
SCIENCES · ENGINEERING · MEDICINE

THE NATIONAL ACADEMIES PRESS
Washington, DC
www.nap.edu

THE NATIONAL ACADEMIES PRESS   500 Fifth Street, NW   Washington, DC 20001

This activity was supported by Contract/Grant No. DE-PI0000010, task DE-DT0009404 between the National Academy of Sciences and the Department of Energy. Any opinions, findings, conclusions, or recommendations expressed in this publication do not necessarily reflect the views of any organization or agency that provided support for the project.

International Standard Book Number-13:   978-0-309-45420-9
International Standard Book Number-10:   0-309-45420-4
Digital Object Identifier: https://doi.org/10.17226/24651

Additional copies of this publication are available for sale from the National Academies Press, 500 Fifth Street, NW, Keck 360, Washington, DC 20001; (800) 624-6242 or (202) 334-3313; http://www.nap.edu.

Printed in the United States of America.

Suggested citation: National Academies of Sciences, Engineering, and Medicine. 2017. *Valuing Climate Damages: Updating Estimation of the Social Cost of Carbon Dioxide*. Washington, DC: The National Academies Press. doi: https://doi.org/10.17226/24651.

The National Academies of
SCIENCES · ENGINEERING · MEDICINE

The **National Academy of Sciences** was established in 1863 by an Act of Congress, signed by President Lincoln, as a private, nongovernmental institution to advise the nation on issues related to science and technology. Members are elected by their peers for outstanding contributions to research. Dr. Marcia McNutt is president.

The **National Academy of Engineering** was established in 1964 under the charter of the National Academy of Sciences to bring the practices of engineering to advising the nation. Members are elected by their peers for extraordinary contributions to engineering. Dr. C. D. Mote, Jr., is president.

The **National Academy of Medicine** (formerly the Institute of Medicine) was established in 1970 under the charter of the National Academy of Sciences to advise the nation on medical and health issues. Members are elected by their peers for distinguished contributions to medicine and health. Dr. Victor J. Dzau is president.

The three Academies work together as the **National Academies of Sciences, Engineering, and Medicine** to provide independent, objective analysis and advice to the nation and conduct other activities to solve complex problems and inform public policy decisions. The National Academies also encourage education and research, recognize outstanding contributions to knowledge, and increase public understanding in matters of science, engineering, and medicine.

Learn more about the National Academies of Sciences, Engineering, and Medicine at **www.national-academies.org**.

The National Academies of
SCIENCES · ENGINEERING · MEDICINE

Reports document the evidence-based consensus of an authoring committee of experts. Reports typically include findings, conclusions, and recommendations based on information gathered by the committee and committee deliberations. Reports are peer reviewed and are approved by the National Academies of Sciences, Engineering, and Medicine.

Proceedings chronicle the presentations and discussions at a workshop, symposium, or other convening event. The statements and opinions contained in proceedings are those of the participants and have not been endorsed by other participants, the planning committee, or the National Academies of Sciences, Engineering, and Medicine.

For information about other products and activities of the National Academies, please visit nationalacademies.org/whatwedo.

vi

Acknowledgments

A number of individuals and organizations contributed to the successful completion of this report. We wish to thank the Interagency Working Group for the Social Cost of Greenhouse Gases for initiating this study and for the study's sponsor, the U.S. Department of Energy, for supporting our work.

Casey Wichman, Resources for the Future, was the study's technical consultant. We wish to thank Casey for the many contributions he made to both Phase 1 and this final report and throughout the course of the study. Casey's expertise and attention to detail improved the quality of both reports.

Over the course of the study, committee members benefited from discussion and presentations by the many individuals who participated in the committee's information-gathering meetings. Appendix B provides a full listing.

Several individuals contributed to the report through commissioned research. We wish to thank Delavane Diaz, Electric Power Research Institute (EPRI), and Frances Moore, Department of Environmental Science and Policy at University of California, Davis, for performing a literature review of climate impacts and damages that was important for Chapter 5. We also wish to thank Bentley Coffey, Sanford School of Public Policy, Duke University, for conducting forecasting studies on long-term growth rates that are described in Appendix D and contributed to Chapter 3. We would also like to thank and recognize Scott Doney, Marine Chemistry and Geochemistry, Woods Hole Oceanographic Institution, for his review of the analysis and calculations used in Chapter 4 and Appendix F.

Thanks are also due to the National Academies of Sciences, Engineering, and Medicine project staff and staff of the Division of Behavioral and Social Sciences and Education (DBASSE). Jennifer (Jenny) Heimberg directed the study and played a key role in project management, report drafting, and the review process. Mary Ghitelman managed the study's logistical and administrative needs, making sure meetings ran efficiently and smoothly. Kirsten Sampson-Snyder guided the report through the National Academies review process, and Eugenia Grohman provided editorial direction. Toby Warden, interim director of the Board on Environmental Change and Society (after November 2016), assisted with the report release, and Jenell Walsh-Thomas, Christine Mirzayan Science and Technology Fellow, stepped in to support final report production activities. Finally, Mary Ellen O'Connell, executive director of DBASSE and interim director of the Board on Environmental Change and Society (through November 2016), helped us from the study's initiation to its completion; we are thankful for her guidance throughout.

This report has been reviewed in draft form by individuals chosen for their diverse perspectives and technical expertise. The purpose of this independent review is to provide candid and critical comments that will assist the institution in making its published report as sound as possible and to ensure that the report meets institutional standards for objectivity, evidence, and responsiveness to the study charge. The review comments and draft manuscript remain confidential to protect the integrity of the deliberative process.

We thank the following individuals for their participation in the review of this report: Hadi Dowlatabadi, Institute for Resources Environment and Sustainability, University of British Columbia; James (Jae) Edmonds, Joint Global Change Research Institute, Pacific Northwest National Laboratory; Karen Fisher-Vanden, Environmental and Resource Economics, The Pennsylvania State University; Michael Greenstone, Energy Policy Institute at Chicago and Department of Economics, University of Chicago; Anthony C. Janetos, The Frederick S. Pardee Center for the Study of the Longer-Range Future, Boston University; Peter B. Kelemen, Department of Earth and Environmental Sciences, Columbia University, and Lamont-Doherty Earth Observatory; Bryan K. Mignone, Corporate Strategic Research, ExxonMobil Research and Engineering Company; Richard H. Moss, Joint Global Change Research Institute, University of Maryland; Elisabeth Moyer, Department of Geophysical Sciences, University of Chicago; Richard L. Revesz, New York University School of Law (emeritus); David A. Weisbach, Law School and Computation Institute, University of Chicago, and Argonne National Laboratories; Jonathan B. Wiener, Law, Environmental Policy, and Public Policy Law School, Nicholas School of the Environment, and Sanford School of Public Policy, Duke University;

and Gary W. Yohe, Economics and Environmental Studies, Wesleyan University.

Although the reviewers listed above have provided many constructive comments and suggestions, they were not asked to endorse the conclusions nor did they see the final draft of the report before its release. The review of this report was overseen by Elisabeth M. Drake, Energy Laboratory emerita, Massachusetts Institute of Technology, and Charles F. Manski, Department of Economics, Northwestern University, who were responsible for making certain that an independent examination of this report was carried out in accordance with institutional procedures and that all review comments were carefully considered. Responsibility for the final content of this report rests entirely with the authoring committee and the institution.

Finally, the dedication, collegiality, and hard work of the committee are especially appreciated. We recognize that the committee members have demanding positions outside of this study. We thank them for their time and commitment to this report.

Maureen L. Cropper, *Cochair*
Richard G. Newell, *Cochair*

Contents

References   **191**

Appendixes

Acronyms

AMOC	Atlantic meridional overturning circulation
AR5	IPCC's Fifth Assessment Report
CDF	cumulative distribution function
CGE	computable general equilibrium
CH_4	methane
CMIP	Coupled Model Intercomparison Project
CMIP5	Coupled Model Intercomparison Project, Phase 5
CO_2	carbon dioxide
CO2SYS.m	carbon systems calculation code
DIC	dissolved inorganic carbon
DICE	Dynamic Integrated Climate-Economy (model)
ECP	extended concentration pathway
ECS	equilibrium climate sensitivity
EMF-22	Energy Modeling Forum's 22nd study
EPPA	emissions prediction and policy analysis
ESMs	Earth system models
FAIR	finite amplitude impulse response
FUND	Framework for Uncertainty, Negotiation and Distribution (model)

GCAM	Global Change Assessment Model
GDP	gross domestic product
GHGs	greenhouse gas
GLODAP2	Global Data Analysis Project, 2
GMSL	global mean sea level
GNI	gross national income
Gt C	gigaton, or 1 billion tons, of carbon
Gt CO_2	gigaton of carbon dioxide
GWP	gross world product
IAMs	integrated assessment models
IIASA	International Institute for Applied Systems Analysis
IPCC	Intergovernmental Panel on Climate Change
IPT	initial pulse adjustment timescale
IWG	Interagency Working Group on the Social Cost of Carbon (through July 2016); Interagency Working Group on the Social Cost Greenhouse Gases (beginning August 2016)
kg/m^3	kilogram per cubic meter
MERGE	model for estimating the regional and global effects of greenhouse gas reductions
microeq/kg	microequivalents per kilogram
micromol/kg	micromoles per kilogram
micromol/$PgCO_2$	micromole per petagram of carbon dioxide
MW	Mueller-Watson
N_2O	nitrous oxide
NPV	net present value
OECD	Organisation for Economic Co-operation and Development
OMB	U.S. Office of Management and Budget
PAGE	Policy Analysis of the Greenhouse Effect (model)
PAGE09	2009 version of PAGE
pCO_2	atmospheric carbon dioxide
POLES	Prospective Outlook on Long-term Energy Systems

ppm	parts per million
PSU	practical salinity unit
RCPs	representative concentration pathways
RCP 2.6, 4.5, 8.5, 6.0	versions of representative concentration pathways
RIAs	federal regulatory impact analyses
RICE	Regional Integrated Climate-Economy (model)
RWF	realized warming fraction (acronym used prior to 2016)
SCC	social cost of carbon (acronym used prior to 2016)
SC-CO$_2$	social cost of carbon dioxide
SC-IAM	social cost of carbon IAMs
SCMs	simple climate models
SESMs	simple Earth system models
SIR	static impulse response
SLR	sea level rise
TCR	transient climate response
TCRE	transient climate response to emissions
W/m^2	watts per square meter
WITCH	World Induced Technical Change Hybrid Model

Executive Summary

The social cost of carbon (SC-CO_2) is an economic metric intended to provide a comprehensive estimate of the net damages—that is, the monetized value of the net impacts, both negative and positive—from the global climate change that results from a small (1 metric ton) increase in carbon dioxide (CO_2) emissions. Under Executive Orders regarding regulatory impact analysis and as required by a court ruling, the U.S. government has since 2008 used estimates of the SC-CO_2 in federal rulemakings to value the costs and benefits associated with changes in CO_2 emissions. In 2010, the Interagency Working Group on the Social Cost of Greenhouse Gases (IWG) developed a methodology for estimating the SC-CO_2 across a range of assumptions about future socioeconomic and physical earth systems.

The IWG asked the National Academies of Sciences, Engineering, and Medicine to examine potential approaches, along with their relative merits and challenges, for a comprehensive update to the current methodology. The task was to ensure that the SC-CO_2 estimates reflect the best available science, focusing on issues related to the choice of models and damage functions, climate science modeling assumptions, socioeconomic and emissions scenarios, presentation of uncertainty, and discounting.

Integrated assessment models (IAMs) are currently used by the IWG to estimate the economic consequences of CO_2 emissions. The IAMs define baseline emission trajectories by projecting future economic growth, population, and technological change. In these IAMs, a 1 metric ton increase in CO_2 emissions is added to the baseline emissions trajectory. This emissions increase is translated into an increase in atmospheric CO_2 concentrations, which results in an increase in global average temperature. This temperature change, as well as changes in other relevant variables, including CO_2 concentrations and income, is translated (either explicitly or implicitly) to physical impacts and monetized damages. These damages include, but are not limited to, market damages, such as changes in net agricultural productivity, energy use, and property damage from increased flood risk, as well as nonmarket damages, such as those to human health and to the services that natural ecosystems provide to society. Because most of the warming caused by an emission of CO_2 into the atmosphere persists for well over a millennium, changes in CO_2 emissions today may affect economic outcomes for centuries to come. Streams of monetized damages over time are converted into present value terms by discounting. The present value of damages reflects society's willingness to trade value in the future for value today.

The IWG methodology combines tens of thousands of SC-CO_2 results obtained from running three IAMs using five different socioeconomic and emissions projections, a common distribution of equilibrium climate sensitivity (a parameter that characterizes the relationship between CO_2 concentrations and long-term global average temperature change), and distributions for other parameters. These results yield three distributions of SC-CO_2 values for three different discount rates, from which the IWG calculated an average value for each discount rate. The IWG's current estimate of the SC-CO_2 in the year 2020 for a 3.0 percent discount rate is $42 per metric ton of CO_2 emissions in 2007 U.S. dollars. If, for example, a particular regulation was projected to reduce CO_2 emissions by 1 million metric tons in 2020, the estimate of the value of its CO_2 emissions benefits in 2020 for this SC-CO_2 would be $42 million dollars.

The Committee on Assessing Approaches to Updating the Social Cost of Carbon recommends near-term improvements to the existing IWG SC-CO_2 estimation methodology, as well as longer-term recommendations for comprehensive updates, and it offers research priorities. Both near- and longer-term recommendations provide guidance to improve the scientific basis, characterization of uncertainty, and transparency of the SC-CO_2 estimation framework within the federal regulatory context for which the SC-CO_2 was developed.

The committee specifies criteria for future updates to the SC-CO_2. It also recommends an integrated modular approach for SC-CO_2 estimation to better satisfy the specified criteria and to draw more readily on expertise from the wide range of scientific disciplines relevant to SC-CO_2 estimation. Under this approach, each step in SC-CO_2 estimation is developed as a module—socioeconomic, climate, damages, and discounting—

that reflects the state of scientific knowledge in the current, peer-reviewed literature.

Because it is important to update estimates as the science and economic understanding of climate change and its impacts improve over time, the committee recommends that estimates of the SC-CO_2 be updated in a three-step process at regular intervals of approximately 5 years. This timing would balance the benefit of incorporating evolving research against the need for a thorough and predictable process.

For each module, the committee recommends near-term changes given the current state of the science. The recommended changes would be feasible to implement in the next 2-3 years and would improve the performance of each part of the analysis with respect to the primary criteria.

- The socioeconomic module should use statistical methods and expert judgment for projecting distributions of economic activity, population growth, and emissions into the future.
- The climate module should use a simple Earth system model that satisfies well-defined diagnostic tests to confirm that it properly captures the relationships between CO_2 emissions, atmospheric CO_2 concentrations, and global mean surface temperature change and sea level rise.
- The damages module should improve and update existing formulations of climate change damages, make calibrations transparent, present disaggregated results, and address correlation between different formulations. This update should draw on recent scientific literature relating to both empirical estimation and process-based modeling of damages.
- The discounting module should incorporate the relationship between economic growth and discounting. The committee also recommends that the IWG provide guidance on how the SC-CO_2 estimates should be combined in regulatory impact analyses with other calculations.

In addition, the committee details longer-term research that could improve each module and incorporate interactions within and feedbacks across modules. These advances will require significant investments in both economic and climate modeling research, particularly research related to the assessment of climate damages and to socioeconomic and emission projections.

Summary

The social cost of carbon ($SC\text{-}CO_2$) for a given year is an estimate, in dollars, of the present discounted value of the future damage caused by a 1 metric ton increase in carbon dioxide (CO_2) emissions into the atmosphere in that year or, equivalently, the benefits of reducing CO_2 emissions by the same amount in that year. The $SC\text{-}CO_2$ is intended to provide a comprehensive measure of the net damages—that is, the monetized value of the net impacts—from global climate change that result from an additional ton of CO_2.[1] Those damages include, but are not limited to, changes in net agricultural productivity, energy use, human health, property damage from increased flood risk, as well as nonmarket damages, such as the services that natural ecosystems provide to society. Many of these damages from CO_2 emissions today will affect economic outcomes throughout the next several centuries. Federal agencies are required to use the $SC\text{-}CO_2$ to value the CO_2 emission reduction benefit of proposed regulations, including emission and fuel economy standards for vehicles; emission standards for industrial manufacturing, power plants, and solid waste incineration; and appliance energy efficiency standards.

The Interagency Working Group on the Social Cost of Greenhouse Gases[2] (IWG) developed a methodology for estimating the $SC\text{-}CO_2$. That

---

[1]Here and throughout this report, "damage" represents the net effects of both the negative and positive economic outcomes of climate change.

[2]Until 2016 the name of the group was the Interagency Working Group on the Social Cost of Carbon.

methodology has been applied to produce estimates that U.S. government agencies use in regulatory impact analyses under Executive Order 12866. The IWG requested the National Academies of Science, Engineering, and Medicine to undertake a study examining potential approaches, along with their relative merits and challenges, for a comprehensive update to the SC-CO$_2$ estimates.

AN INTEGRATED MODULAR APPROACH

The committee's conclusions and recommendations highlight four components of analysis or "modules" involved in estimating the SC-CO$_2$—socioeconomic and emissions projections, climate modeling, estimation of climate impacts and damages, and discounting net monetary damages. Each module comprises conceptual formulations and theory, computer models, and other analytical frameworks; each is supported by its own specialized disciplinary expertise. The SC-CO$_2$ estimation framework put forward by the committee integrates these four modules, and, when possible, taking into account the interdependencies among them.

Current estimates of the SC-CO$_2$ are obtained by pooling estimates of monetized damages produced by three reduced-form integrated assessment models (IAMs) that feature prominently in the literature assessing the benefits and costs of climate change mitigation: the committee refers to these as SC-IAMs. Each SC-IAM contains its own modeling components along the lines of the four modules described. The IWG ran each SC-IAM with a common set of socioeconomic scenarios and a common distribution of the equilibrium climate sensitivity, as well as model-specific distributions for other parameters.

> **CONCLUSION 2-1** For at least some steps in the SC-CO$_2$ estimation framework, using a common module—rather than averaging the results from multiple models—can improve transparency and consistency of key assumptions with the peer-reviewed science and can improve control over the uncertainty representation, including structural uncertainty. This rationale underlies the Interagency Working Group's use of the same socioeconomic scenarios, discount rates, and distribution for climate sensitivity across IAMs, as well as the committee's suggestion in its Phase 1 report that the IWG develop or adopt a common climate module.

> **CONCLUSION 2-2** An integrated modular framework for SC-CO$_2$ estimation can provide a transparent identification of the inputs, outputs, uncertainties, and linkages among the dif-

ferent steps of the SC-CO$_2$ estimation process. This framework can also provide a mechanism for incorporating new scientific evidence and for facilitating regular improvement of the framework modules and resulting estimates by engaging experts across the varied disciplines that are relevant to each module.

RECOMMENDATION 2-1 The Interagency Working Group should support the creation of an integrated modular SC-CO$_2$ framework that provides a transparent articulation of the inputs, outputs, uncertainties, and linkages among the different steps of SC-CO$_2$ estimation. For some modules within this framework, the best course of action may be for the government to develop a new module, while for other modules the best course of action may be to adapt one or more existing models developed by the scientific community.

RECOMMENDATION 2-2 The Interagency Working Group should use three criteria to evaluate the overall integrated SC-CO$_2$ framework and the modules to be used in that framework: scientific basis, uncertainty characterization, and transparency.

- Scientific basis: Modules, their components, their interactions, and their implementation should be consistent with the state of scientific knowledge as reflected in the body of current, peer-reviewed literature.
- Uncertainty characterization: Key uncertainties and sensitivities, including functional form, parameter assumptions, and data inputs, should be adequately identified and represented in each module. Uncertainties that cannot be or have not been quantified should be identified.
- Transparency: Documentation and presentation of results should be adequate for the scientific community to understand and assess the modules. Documentation should explain and justify design choices, including such features as model structure, functional form, parameter assumptions, and data inputs, as well as how multiple lines of evidence are combined. The extent to which features are evidence-based or judgment-based should be explicit. Model code should be available for review, use, and modification by researchers.

In the integrated modular CO$_2$ framework, the first of the four modules would generate estimates of future population and gross domestic product (GDP). From these, it would generate projections of greenhouse gas emissions. Each emissions path would serve as a baseline to which an emission pulse is added in order to estimate the incremental impact of an additional ton of CO_2 released in a particular year. Given projected emissions, the climate module would generate estimates of CO_2 concentrations in the atmosphere and ocean, surface temperature change, and sea level rise. Together with the associated population and GDP projections, these climate results would serve as inputs to the damages module that would calculate the monetary value, each year, of net climate damages due to projected emissions.

Each of these modules would include data inputs or structural elements that are treated as uncertain, leading to outputs in the form of distributions of estimates for each year rather than a single value. The discounting module would sum the future stream of monetized damage estimates to a single present value for each of the possible future "states of the world" that are embodied in the analysis in the earlier steps in the SC-CO_2 estimation process.

In addition to recommendations regarding the incorporation of uncertainty in the modeling process, the committee reiterates its recommendations on the presentation of uncertainty from its Phase 1 report. Specifically, it is important that the sources of uncertainty in SC-CO_2 estimation be made clear. In future updates to the technical support documentation of the SC-CO_2 estimates, a discussion of various types of uncertainty and how they are handled in estimating the SC-CO_2, as well as sources of uncertainty that are not captured in current SC-CO_2 estimates, would improve transparency.

The main disadvantage of a focus on individual modules is the potential neglect of important feedbacks between components of the system. Successful implementation of a modular framework in the longer term will require attention to the interactions among the modules, and modification of the overall structure to incorporate findings and approaches from ongoing research on the human-environment-climate system.

RECOMMENDATION 2-3 The Interagency Working Group should continue to monitor research that identifies and explores the magnitude of various interactions and feedbacks in the human-climate system including those not represented in implementation of the proposed modular SC-CO_2 estimation framework. The IWG should include interactions and feedbacks among the modules of the SC-CO_2 framework if they are found to significantly affect SC-CO_2 estimates.

Due to the global nature of the impacts that result from CO_2 emissions regardless of where they originate, efforts to estimate the SC-CO_2 by both the scientific community and the IWG have focused on total global damages, rather than the damages to an individual country such as the United States. At the same time, the IWG recognized that this approach "represents a departure from past practices, which tended to put greater emphasis on a domestic measure of SC-CO_2 (limited to impacts of climate change experienced within U.S. borders)" (Interagency Working Group on the Social Cost of Carbon, 2010, p. 10). The IWG therefore provided rough estimates of the proportion of global damages attributable directly to impacts within U.S. borders.

Accurately estimating the damage of CO_2 emissions for the United States involves more than examining the direct impacts of climate change that occur within U.S. physical borders. The IWG has noted that climate change in other regions of the world could affect the United States, through such pathways as global migration, economic destabilization, and political destabilization. In addition, the United States may be affected by changes in the economic conditions of its trading partners. The current SC-IAMs do not fully account for these types of interactions. The implications of U.S. emissions or mitigation thereof on levering actions by other countries is another consideration affecting the accurate estimation of the domestic, relative to the global, damages from U.S. CO_2 emissions.

CONCLUSION 2-4 Estimation of the net damages per ton of CO_2 emissions to the United States alone, beyond the approximations done by the IWG, is feasible in principle; however it is limited in practice by the existing SC-IAM methodologies, which focus primarily on global estimates and do not model all relevant interactions among regions. It is important to consider what constitutes a domestic impact in the case of a global pollutant that could have international implications that impact the United States. More thoroughly estimating a domestic SC-CO_2 would therefore need to consider the potential implications of climate impacts on, and actions by, other countries, which also have impacts on the United States.

The committee recommends a regularized process for updating SC-CO_2 estimates to enhance their scientific credibility and provide a way for experts to suggest both improvements for updates and priorities for research.

RECOMMENDATION 2-4 The Interagency Working Group should establish a regularized three-step process for updating

the SC-CO$_2$ estimates. An update cycle of roughly 5 years would balance the benefit of responding to evolving research with the need for a thorough and predictable process. In the first step, the interagency process and associated technical efforts should draw on internal and external technical expertise and incorporate scientific peer review. In the second step, draft revisions to the SC-CO$_2$ methods and estimates should be subject to public notice and comment, allowing input and review from a broader set of stakeholders, the scientific community, and the public. In the third step, the government's approach to estimating the SC-CO$_2$ should be regularly reviewed by an independent scientific assessment panel to identify improvements for potential future updates and research needs.

SOCIOECONOMIC MODULE

The purpose of a socioeconomic module is to provide a set of projections of population and GDP, which in turn drive projections of CO$_2$ and other relevant climate-forcing emissions that are inputs to the calculation of a baseline climate trajectory. The baseline trajectory influences the response of the climate to a pulse of CO$_2$ emissions. Estimates of population and GDP, possibly disaggregated by region and sector, are also direct inputs to the estimation of climate damages, and the trajectory of GDP per capita also feeds into the recommended discounting procedure.

RECOMMENDATION 3-1 In addition to applying the committee's overall criteria for scientific basis, uncertainty characterization, and transparency (see Recommendation 2-2 in Chapter 2), the Interagency Working Group should evaluate potential socioeconomic modules according to four criteria: time horizon, future policies, disaggregation, and feedbacks.

- Time horizon: The socioeconomic projections should extend far enough in the future to provide inputs for estimation of the vast majority of discounted climate damages.
- Future policies: Projections of emissions of CO$_2$ and other important forcing agents should take account of the likelihood of future emissions mitigation policies and technological developments.
- Disaggregation: The projections should provide the sectoral and regional detail in population and GDP necessary for damage calculations.

- Feedbacks: To the extent possible, the socioeconomic module should incorporate feedbacks from the climate and damages modules that have a significant impact on population, GDP, or emissions.

To produce a module satisfying the criteria in Recommendation 3-1, the committee offers a recommendation for the near term.

RECOMMENDATION 3-2 In the near term, to develop a socioeconomic module and projections over the relevant time horizon, the Interagency Working Group should:

- Use an appropriate statistical technique to estimate a probability density of average annual growth rates of global per capita GDP. Choose a small number of values of the average annual growth rate to represent the estimated density. Elicit expert opinion on the desirability of possible modifications to the implied projections of per capita GDP, particularly after 2100.
- Work with demographers who have produced probabilistic projections through 2100 to create a small number of population projections beyond 2100 to represent a probability density function. Development of such projections should include both the extension of existing statistical models and the elicitation of expert opinion for validation and adjustment, particularly after 2100. Should either the economic or demographic experts suggest that correlation between economic and population projections is important, this could be included.
- Use expert elicitation, guided by information on historical trends and emissions consistent with different climate outcomes, to produce a small number of emissions trajectories for each forcing agent of interest conditional on population and income scenarios.
- Develop projections of sectoral and regional GDP and regional population using scenario libraries, published regional or national population projections, detailed-structure economic models, SC-IAMs, or other sources.

In the longer term, there are many advantages to investing in the construction of a dedicated socioeconomic projection framework designed for the task. Existing detailed-structure models were formulated to meet objectives different from those of the IWG.

RECOMMENDATION 3-3 In the longer term, the Interagency Working Group should engage in the development of a new socioeconomic module, based on a detailed-structure model, that meets the criteria of scientific basis, uncertainty characterization, and transparency, is consistent with the best available judgment regarding the probability distributions of uncertain parameters and that has the following characteristics:

- provides internally consistent probabilistic projections, consistent with elicited expert opinion, as far beyond 2100 as required to capture the vast majority of discounted damages, taking into account the increased uncertainty regarding technology, policies, and social and economic structures in the distant future;
- provides probabilistic regional and sectoral projections consistent with requirements of the damage module, taking into account historical experience, expert judgment, and increasing uncertainty over time regarding the regional and sectoral structure of the global economy;
- captures important feedbacks from the climate and damage modules that affect capital stocks, productivity, and other determinants of socioeconomic and emissions projections. It should enable interactions among the modules to ensure consistency among economic growth, emissions, and their consequences; and
- is developed in conjunction with the climate and damage modules, to provide a coherent and manageable means of propagating uncertainty through the components of the SC-CO_2 estimation procedure.

Development of such a framework, designed to satisfy the long-term needs of SC-CO_2 estimation, would represent an advance in economic modeling. Chapter 7 includes a set of conclusions about the research needed on economic modeling frameworks and model development for the long term.

CLIMATE MODULE

The purpose of a climate module is to take outputs of the socioeconomic module (particularly emissions of CO_2 and other climate forcing agents) and estimate their effect on physical climate variables, such as a time series of temperature change, at the spatial and temporal resolution required by the damages module. Thus, it must translate greenhouse gas emissions into atmospheric concentrations, translate concentrations of CO_2 and other climate forcers into their radiative effects ("forcing"), translate forcing into global mean surface temperature response, and generate other climatic variables that may be needed by the damage module. In so doing, it must accurately represent within a probabilistic context the current understanding of the climate and carbon cycle systems and associated uncertainties.

A simple Earth system model would be appropriate for the SC-CO_2 setting, and it is important that such a model be considered for use in SC-CO_2 calculations. Such a model would reflect current scientific understanding of the relationships between greenhouse gas emissions, concentrations, radiative forcing, and global mean surface temperature change, as well as their uncertainty and profiles over time.

RECOMMENDATION 4-1 In the near term, the Interagency Working Group should adopt or develop a climate module that captures the relationships between CO_2 emissions, atmospheric CO_2 concentrations, and global mean surface temperature change, as well as their uncertainty, and projects their profiles over time. The module should apply the overall criteria for scientific basis, uncertainty characterization, and transparency (see Recommendation 2-2 in Chapter 2). In the context of the climate module, this means:

- **Scientific basis and uncertainty characterization: The module's behavior should be consistent with the current, peer-reviewed scientific understanding of the relationships over time between CO_2 emissions, atmospheric CO_2 concentrations, and CO_2-induced global mean surface temperature change, including their uncertainty. The module should be assessed on the basis of its response to long-term forcing trajectories (specifically, trajectories designed to assess equilibrium climate sensitivity, transient climate response and transient climate response to emissions, as well as historical and high- and low-emissions scenarios) and its response to a pulse of CO_2 emissions. The assessment of the module should be formally documented.**
- **Transparency and simplicity: The module should strive for transparency and simplicity so that the central tendency and range of uncertainty in its behavior are readily understood, reproducible, and amenable to improvement over time through the incorporation of evolving scientific evidence.**

The climate module should also meet the following additional criterion:

- Incorporation of non-CO_2 forcing: The module should be formulated such that effects of non-CO_2 forcing agents can be incorporated, which will allow both for more accurate reflection of baseline trajectories and for the same model to be used to assess the social cost of non-CO_2 forcing agents in a manner consistent with estimates of the SC-CO_2.

RECOMMENDATION 4-2 To the extent possible, the Interagency Working Group should use formal assessments that draw on multiple lines of evidence and a broad body of scientific work, such as the assessment reports of the Intergovernmental Panel on Climate Change, which provide the most reliable estimates of the ranges of key metrics of climate system behavior. If such assessments are not available, the IWG should derive estimates from a review of the peer-reviewed literature, with care taken so as to not introduce inconsistencies with the formally assessed parameters. The assessments should provide ranges with associated likelihood statements and specify complete probability distributions. If multiple interpretations are possible, the selected approach should be clearly described and justified.

An example of a simple Earth system model that satisfies the criteria set forth above is the Finite Amplitude Impulse Response (FAIR) model (see Chapter 4). FAIR includes a minor modification of the model used in the Intergovernmental Panel on Climate Change Fifth Assessment Report to assess the global warming potential of different gases. The committee notes that none of the current SC-IAM climate components fully satisfies the criteria above.

Global mean sea level rise is another key physical variable relevant for estimating climate damages. Global mean sea level rise results from both the transfer of water mass from continental ice sheets and glaciers into the ocean, and also from the volumetric expansion of ocean water as it warms.

RECOMMENDATION 4-3 In the near term, the Interagency Working Group should adopt or develop a sea level rise component in the climate module that (1) accounts for uncertainty in the translation of global mean temperature to global mean sea level rise and (2) is consistent with sea level rise projections available in the literature for similar forcing and temperature pathways. Existing semi-empirical sea level models provide one basis for doing this. In the longer term, research will be necessary to incorporate recent scientific discoveries regarding ice sheet stability in such models.

CO_2 dissolves in seawater to form carbonic acid. As the oceans have absorbed about one-quarter to one-third of anthropogenic CO_2 emissions, the oceans have steadily become more acidic. Modeling of the consequences of ocean acidification is at an early stage, and is mainly carried out using Earth system or regional ocean models with comparable complexity.

RECOMMENDATION 4-4 The Interagency Working Group should adopt or develop a surface ocean pH component within the climate module that (1) is consistent with carbon uptake in the climate module, (2) accounts for uncertainty in the translation of global mean surface temperature and carbon uptake to surface ocean pH, and (3) is consistent with observations and projections of surface ocean pH available in the current peer-reviewed literature. For example, surface ocean pH can be derived from global mean surface temperature and global cumulative carbon uptake using relationships calibrated to the results of explicit models of carbonate chemistry of the surface ocean.

Simple Earth system models produce climate projections that are highly aggregated both spatially and temporally. For example, the FAIR model produces projections of climatological (multi-decadal-average) global mean temperature. Yet people do not live under 30-year global mean conditions. Rather, damages are caused by the day-to-day, place-specific effects of the weather, the statistical properties of which are described by the climate. The damages module will therefore either require geographically and temporally disaggregated climate variables as inputs or such disaggregation will need to occur in the calibration of the relationship between highly aggregated climate variables and resulting damages. The most straightforward approach to transforming global mean variables into more spatially disaggregated variables is to estimate linear relationships between local climate variables (e.g., temperature, precipitation) and global mean temperature, known as pattern scaling.

RECOMMENDATION 4-5 To the extent needed by the damages module, the Interagency Working Group should use disaggre-

gation methods that reflect relationships between global mean quantities and disaggregated variables, such as regional mean temperature, mean precipitation, and frequency of extremes, that are inferred from up-to-date observational data and more comprehensive climate models.

CONCLUSION 4-3 In the near term, linear pattern scaling, although subject to numerous limitations, provides an acceptable approach to estimating some regionally disaggregated variables from global mean temperature and global mean sea level. If necessary, projections based on pattern scaling can be augmented with high-frequency variability estimated from observational data or from model projections. In the longer term, it would be worthwhile to consider incorporating the dependence of disaggregated variables on spatial patterns of forcing, the temporal evolution of patterns under stable or decreasing forcing, and nonlinearities in the relationship between global mean variables and regional variables.

Research focused on improving the representation of the Earth system in the context of coupled climate-economic analyses would improve the reliability of estimates of the SC-CO_2. A list of research topics needed to reach such a goal is outlined in Chapter 7.

DAMAGES MODULE

The purpose of the damages module is to translate a time series of socioeconomic variables (e.g., income and population) and physical climatic variables (e.g., changes in temperature and sea level) into estimates of physical impacts and, when possible, monetized damages over time. To do so, it must represent relationships among physical variables, socioeconomic variables, and damages. The SC-IAMs include damage representations that are either simple and global (e.g., global damages as a function of global mean temperature and gross world product), or are sectorally and regionally disaggregated (e.g., agricultural damages as a function of regional temperature, precipitation change, CO_2 concentrations, and the economic value added or GDP of relevant sectors or regions).

RECOMMENDATION 5-1 In the near term, the Interagency Working Group should develop a damages module using elements from the current SC-IAM damage components and scientific literature. The damages module should meet the committee's overall criteria for scientific basis, transparency, and uncertainty characterization (see Recommendation 2-2, in Chapter 2) and include the following four additional improvements:

1. Individual sectoral damage functions should be updated as feasible.
2. Damage function calibrations should be transparently and quantitatively characterized.
3. If multiple damage formulations are used, they should recognize any correlations between formulations.
4. A summary should be provided of disaggregated (incremental and total) damage projections underlying SC-CO$_2$ calculations, including how they scale with temperature, income, and population.

RECOMMENDATION 5-2 In the longer term, the IWG should develop a damages module that meets the overall criteria for scientific basis, transparency, and uncertainty characterization (see Recommendation 2-2, in Chapter 2) and has the following five features:

1. It should disaggregate market and nonmarket climate damages by region and sector, with results that are presented in both monetary and natural units and that are consistent with empirical and structural economic studies of sectoral impacts and damages.
2. It should include representation of important interactions and spillovers among regions and sectors, as well as feedbacks to other modules.
3. It should explicitly recognize and consider damages that affect welfare either directly or through changes to consumption, capital stocks (physical, human, natural), or through other channels.
4. It should include representation of adaptation to climate change and the costs of adaptation.
5. It should include representation of nongradual damages, such as those associated with critical climatic or socioeconomic thresholds.

DISCOUNTING MODULE

The purpose of a discounting module is to integrate the future stream of monetized damage estimates into a single present value for each state of the world generated by the earlier steps of the SC-CO$_2$ estimation process. Discounting is the procedure by which costs and benefits in future years are made comparable with costs and benefits incurred today. The discount rate refers to a reduction (or "discount") in value that a future cost or benefit is adjusted for each year in the future to be compared with a current cost or benefit. Because the impacts of CO_2 emissions in any particular year persist for many years, the value of avoiding those impacts depends on how much society discounts those future impacts. Due to the power of compounding, small differences in the discount rate can have large impacts on the estimated SC-CO_2.

> **CONCLUSION 6-1** In the current approach of the Interagency Working Group, uncertainty about future discount rates motivates the use of both a lower 2.5 percent rate and higher 5.0 percent rate, relative to the central 3.0 percent rate. However, this approach does not incorporate an explicit connection between discounting and consumption growth that arises under a more structural (e.g., Ramsey-like) approach to discounting. Such an explicit analytic connection is especially important when considering uncertain climate damages that are positively or negatively associated with the level of consumption. The Ramsey formula provides a feasible and conceptually sound framework for modeling the relationship between economic growth and discounting uncertainty.

In formulating its recommendations, the committee makes use of the Ramsey discounting formula, in which the discount rate equals the sum of the pure rate of time preference (δ) and the product of the value of an additional dollar as society grows wealthier (η) and the growth rate of per capita consumption (g).

> **RECOMMENDATION 6-1** The Interagency Working Group should develop a discounting module that explicitly recognizes the uncertainty surrounding discount rates over long time horizons, its connection to uncertainty in economic growth, and, in turn, to climate damages. This uncertainty should be modeled using a Ramsey-like formula, $r = \delta + \eta \cdot g$, where the uncertain discount rate r is defined by parameters δ and η and uncertain per capita economic growth g. When applied to a set of projected damage estimates that vary in their assumptions about per capita economic growth, each projection should use a path of discount rates based on its particular path of per capita economic growth. These discounted damage estimates can then be

used to calculate an average SC-CO$_2$ and an uncertainty distribution for the SC-CO$_2$, conditional on the assumed parameters.

To choose the parameters of a Ramsey-like approach, one could examine empirical assessments of pure time preference and utility curvature or one could choose those parameters to match empirical features of observed interest rates and the long-term relationship between interest rates and economic growth.

RECOMMENDATION 6-2 The Interagency Working Group should choose parameters for the Ramsey formula that are consistent with theory and evidence and that produce certainty-equivalent discount rates consistent, over the next several decades, with consumption rates of interest. The IWG should use three sets of Ramsey parameters, generating a low, central, and high certainty-equivalent near-term discount rate, and three means and ranges of SC-CO$_2$ estimates.

In the regulatory impact analyses required under federal rules, the rate at which future benefits and costs are discounted can significantly alter the estimated present value of the net benefits of that rule. In accordance with guidance from the U.S. Office of Management and Budget, agencies have generally used sensitivity analysis with discount rates of 3.0 and 7.0 percent. The 7.0 percent rate is intended to represent the average before-tax rate of return to private capital in the U.S. economy. The 3.0 percent rate is intended to reflect the rate at which society discounts future consumption, which is particularly relevant if a regulation is expected to affect private consumption directly. Due to the atypically long time frame and important intergenerational consequences associated with CO$_2$ emissions, the IWG has used alternative discount rates for the SC-CO$_2$ of 2.5, 3.0, and 5.0 percent.

Incorporating estimates of the SC-CO$_2$ in a regulatory impact analysis can present a challenge if the SC-CO$_2$ calculation uses discount rates that are different from those used for other benefits and costs in the analysis (e.g., short-term air quality impacts).

RECOMMENDATION 6-3 The Interagency Working Group should be explicit about how the SC-CO$_2$ estimates should be combined in regulatory impact analyses with other cost and benefit estimates that may use different discount rates.

1

Introduction

A variety of rules and regulations considered by the U.S. federal government—such as energy efficiency standards, fuel economy standards, and power plant regulations—affect the emissions of carbon dioxide (CO_2) and other greenhouse gases into the atmosphere.[1] For more than three decades, presidential Executive Orders (EOs) have required that federal agencies consider the monetized impact of effects when conducting regulatory impact analyses: see Box 1-1. This report takes a pragmatic approach in offering conclusions and recommendations that are consistent with this approach to regulatory analysis.

In 2008, a ruling by the Ninth Circuit Court of Appeals remanded a fuel economy rule to the Department of Transportation, concluding that it was "arbitrary and capricious" to not monetize the benefits of the CO_2 emission reductions in the rule's regulatory impact analysis.[2] In 2009 an interagency working group was formed and developed an approach for estimating the "social cost of carbon" that has been used in dozens of benefit-cost analyses since 2010. The social cost of carbon (SC-CO_2)[3] is

---

[1] A recent Congressional Research Service report, *Federal Citations to the Social Cost of Greenhouse Gases*, includes a table that lists federal actions that used the SC-CO_2 estimates; the earliest action is April 2008. The report is available at https://fas.org/sgp/crs/misc/R44657.pdf [December 2016].

[2] Center for Biological Diversity v. National Highway Traffic Safety Administration, U.S. Court of Appeals, Ninth Circuit, 538 F.3d 1172 (9th Cir. 2008).

[3] The acronym for the social cost of carbon in the committee's interim report was "SCC," following the then-standard acronym. In late August 2016, the newly renamed Interagency

---

BOX 1-1
Regulatory Impact Analysis under Executive Order (EO) 12866

Executive Orders (EOs) 12291 (from 1981 to 1993) and 12866[a] (since 1993) have required that agencies undertake quantitative regulatory impact analysis of regulatory actions, employing a regulatory philosophy based on maximizing the expected net benefits of those actions. As stated in EO 12866 Section 1(a):

> In deciding whether and how to regulate, agencies should assess all costs and benefits of available regulatory alternatives, including the alternative of not regulating. Costs and benefits shall be understood to include both quantifiable measures (to the fullest extent that these can be usefully estimated) and qualitative measures of costs and benefits that are difficult to quantify, but nevertheless essential to consider. Further, in choosing among alternative regulatory approaches, agencies should select those approaches that maximize net benefits. . . .

Since 2003, U.S. Office of Management and Budget (OMB) Circular A-4[b] has provided specific guidance for conducting regulatory impact analysis under EO 12866, replacing prior guidance. With respect to the treatment of uncertainty in regulatory impact analysis, Circular A-4 provides additional context for the efforts of this committee, stating (p. 18):

> When benefit and cost estimates are uncertain . . . you should report benefit and cost estimates (including benefits of risk reductions) that reflect

---

defined for a given year as the present discounted value of the future damage[4] caused by a 1 metric ton increase in CO_2 emissions to the atmo-

---

Working Group on the Social Cost of Greenhouse Gases (previously, the Interagency Working Group on the Social Cost of Carbon) introduced the acronym "$SC\text{-}CO_2$." This report uses the new acronym, except when referring to text from previously published documents.

[4]Throughout this report, "damage" represents the net effects of both negative and positive economic impacts of climate change. When incorporated in a benefit-cost analysis, such as a regulatory impact analysis, these net damages are reflected as a benefit of emissions reduction. In benefit-cost analysis, the benefit of a commodity is measured by what people are willing to pay for it. It is important to note that willingness to pay is constrained by ability to pay. The notion that the value attached to a commodity should be constrained by resources is fundamental to economics. When one values the output of a commodity sold in markets one uses a demand function, which reflects the willingness to pay of consumers for purchasing additional units of the good. This is conditional on the distribution of income in society. It is when costs and benefits are added together to determine the net benefits of a decision that principles of benefit-cost analysis enter in. In measuring the economic net benefits of an action, one compares the benefits, as defined above, with the costs of the action. This is an appropriate decision-making criterion, but it does not directly take distributional issues

the full probability distribution of potential consequences. Where possible, present probability distributions of benefits and costs and include the upper and lower bound estimates as complements to central tendency and other estimates.

If fundamental scientific disagreement or lack of knowledge prevents construction of a scientifically defensible probability distribution, you should describe benefits or costs under plausible scenarios and characterize the evidence and assumptions underlying each alternative scenario.

Circular A-4 elaborates further (p. 41) that regulatory impact analysis should:

Apply a formal probabilistic analysis of the relevant uncertainties—possibly using simulation models and/or expert judgment,...[which]...combined with other sources of data, can be combined in Monte Carlo simulations to derive a probability distribution of benefits and costs.

---

[a]For the text of EO 12866, see https://www.archives.gov/files/federal-register/executive-orders/pdf/12866.pdf [January 2017]. Another order released in 2011, EO 13563, which reaffirmed and supplemented EO 12866, directs agencies to conduct regulatory actions based on the best available science and to use best available techniques to quantify benefits or costs as accurately as possible. See https://www.gpo.gov/fdsys/pkg/FR-2011-01-21/pdf/2011-1385.pdf [November 2016].

[b]For the text of Circular A-4, see https://obamawhitehouse.archives.gov/sites/default/files/omb/assets/omb/circulars/a004/a-4.pdf [January 2017].

sphere in that year, or, equivalently, the benefits of reducing CO_2 emissions by the same amount in that year.

The Interagency Working Group on the Social Cost of Greenhouse Gases (IWG),[5] composed of experts from multiple federal agencies, develops and maintains the SC-CO_2 estimates. The current estimation approach

---

into account. Some individuals may face net costs from the action, and others may face net gains. To provide information on distributional impacts for policy making, the U.S. Office of Management and Budget (OMB) guidance suggests that the distribution of costs and benefits be measured in regulatory impact analysis. Distributional effects can also be reflected in benefit-cost analysis using welfare weights, although this is rarely done in practice and is not permitted in regulatory impact analysis.

[5]The IWG is cochaired by the Council of Economic Advisors and OMB; the Office of Management and Budget; the other members are the Council on Environmental Quality, the Domestic Policy Council, the Department of Agriculture, the Department of Commerce, the Department of Energy, the Department of the Interior, the Department of Transportation, the Department of the Treasury, the Environmental Protection Agency, the National Economic Council, and the Office of Science and Technology Policy.

was developed in 2009-2010 and released in 2010. The approach has not changed since this initial release, although individual model modifications and other changes were made in 2013 and 2015. The IWG is considering more significant updates to the approach used to estimate the SC-CO$_2$ and asked the National Academies of Sciences, Engineering, and Medicine (hereafter referred to as the Academies) to make recommendations on potential approaches that warrant consideration in future updates of the SC-CO$_2$ estimates. The charge to the Academies also requested recommendation for research that would advance the science in areas that are particularly useful for estimating the SC-CO$_2$. See Box 1-2 for the full statement of task for the committee.

The committee interpreted the charge as focusing specifically on the SC-CO$_2$ for its use in federal regulatory impact analysis. As discussed above, the committee therefore developed its conclusions and recommendations to be consistent with an overall analytical approach based on the computation of expected net present value, taking into account that

---

BOX 1-2
Statement of Task

An ad hoc multi-disciplinary committee will be appointed to inform future revisions to estimates of the social cost of carbon (SCC) developed and used by the federal government. The committee will examine the merits and challenges of potential approaches for both a near-term limited update and longer-term comprehensive updates to ensure that the SCC estimates continue to reflect the best available science and methods. The study will be conducted in two phases and will result in two reports.

Phase 1. In phase 1, the committee will assess the technical merits and challenges of a narrowly focused update to the SCC estimates and make a recommendation on whether to conduct an update of the SCC estimates prior to recommendations related to a more comprehensive update based on its review of the science related to the topics covered in the second phase. Specifically, the committee will consider whether an update is warranted based on the following:

1. Updating the probability distribution for the equilibrium climate sensitivity (ECS) to reflect the recent consensus statement in the Fifth Assessment Report of the Intergovernmental Panel on Climate Change (IPCC), rather than the current calibration used in the SCC estimates, which were based on the most authoritative scientific consensus statement available at the time (the 2007 Fourth IPCC Assessment).
2. Recalibrating the distributional forms for the ECS by methods other than the currently-used Roe and Baker (2007) distribution.

the SC-CO$_2$ is one of a large number of variables that enter into a typical regulatory impact analysis. In doing so, the committee notes that the particular regulations of interest are typically of only incremental impact in the context of total U.S. or global CO$_2$ emissions. The resulting SC-CO$_2$ estimates are therefore not necessarily applicable for use as the basis of very large-scale policy issues, such as a comprehensive national carbon price. At a minimum, care needs to be taken in such applications of the SC-CO$_2$.

The IWG's formulation of the SC-CO$_2$ also differs from much academic work on the issue, which often focuses on optimal global CO$_2$ control: in this work, an optimal emissions control level is set so that its marginal cost is equal to marginal damage, and an SC-CO$_2$ estimate in this case is computed using the optimal emissions pathway. The committee also notes that the analytical framework used in developing the IWG SC-CO$_2$ estimates is based on probability-weighted present value. Although this is appropriate for its application in regulatory impact analysis, it is not

3. Enhancing the qualitative characterization of uncertainties associated with the current SCC estimates in the short-term to increase the transparency associated with using these estimates in regulatory impact analyses. Noting that as part of a potential comprehensive update Part 2 of the charge requests information regarding the opportunity for a more comprehensive, and possibly more formal or quantitative, treatment of uncertainty.

The phase 1 report will be an interim letter report to be completed in 6 months.

Phase 2. In phase 2, which represents the bulk of the statement of task, the committee will examine potential approaches, along with their relative merits and challenges, for a more comprehensive update to the SCC estimates to ensure the estimates continue to reflect the best available science. The Committee will be asked to consider issues related to:

1. an assessment of the available science and how it would impact the choice of integrated assessment models and damage functions;
2. climate science modeling assumptions;
3. socio-economic and emissions scenarios;
4. presentation of uncertainty; and
5. discounting.

Within these areas, the committee will make recommendations on potential approaches that warrant consideration in future updates of the SCC estimates, as well as research recommendations based on their review that would advance the science in areas that are particularly useful for estimating the SCC.

the only framework relevant to decision making under uncertainty in the context of national and international climate policy. Approaches to the treatment of uncertainty are discussed in Chapter 2.

HISTORY AND DEVELOPMENT OF THE SOCIAL COST OF CARBON FOR REGULATORY IMPACT ANALYSIS

Academic research into the estimation of the social costs of greenhouse gas emissions began with work by economist William Nordhaus in the early 1980s (Nordhaus, 1982) and was continued by numerous researchers in the early 1990s (e.g., Ayres and Walter, 1991; Nordhaus, 1991; Haraden, 1992; Peck and Teisberg, 1992; Reilly and Richards, 1993; Fankhauser, 1994).[6] Researchers continued to explore the SC-CO_2 over the subsequent two decades. This research base informed the initial estimates and the current approach adopted by the IWG.[7]

Prior to 2008, changes in CO_2 emissions associated with proposed policies were generally not valued in federal regulatory impact analyses (RIAs). As noted earlier, following a 2008 court ruling, federal agencies began to account for the impact of CO_2 emissions in their analyses. Agencies estimated dollar values for the SC-CO_2 using a variety of methodologies.

In 2009, the Obama Administration formed the IWG and charged it with developing a consistent set of SC-CO_2 estimates to be used in regulatory impact analyses. The IWG comprised relevant subject-matter experts from federal agencies; all federal agencies were welcome to participate. For developing the SC-CO_2 estimates and making decisions on updates, the IWG used consensus-based decision making, relied on existing academic literature and models, and took steps to disclose limitations and incorporate new information (U.S. Government Accountability Office, 2014).[8]

The IWG initially established interim SC-CO_2 values using estimates obtained from the existing literature. These interim values were first used by the U.S. Department of Energy in an RIA for an energy efficiency standard for beverage vending machines in August 2009 (74 *Federal Register* 44914). The IWG continued working on a more in-depth process to estimate the SC-CO_2. In February 2010 it published a set of SC-CO_2 estimates

---

[6]A 2010 report provides an overview of the history of the literature on estimating economic damages due to CO_2 emissions (see National Research Council, 2010, Ch. 5).

[7]The Intergovernmental Panel on Climate Change Fifth Assessment Report provides a database summarizing academic studies on the estimates of the welfare impact of climate change from 1982 to 2012 (Arent et al., 2014).

[8]The organizational process used by the IWG to develop the SC-CO_2 estimates was reviewed by the U.S. Government Accountability Office (2014).

for the years 2010 through 2050 and described the technical methodology for estimation in a *Technical Support Document*.[9] The methodology used the three most widely cited integrated assessment models (IAMs) that are used in benefit-cost analysis of climate policy to produce estimates of the SC-CO_2.[10] This report refers to those models as SC-IAMs.

Four updates to the *Technical Support Documents* related to the SC-CO_2 estimates have occurred since the 2010 release: two in 2013 and one each in 2015 and 2016. None of the updates changed the fundamental methodology used to construct the 2010 SC-CO_2 estimates.

SUMMARY OF THE IWG'S APPROACH TO ESTIMATING THE SOCIAL COST OF CARBON

The technical methodology for constructing the official U.S. SC-CO_2 estimates is discussed in detail in the IWG *Technical Support Documents* (Interagency Working Group on the Social Cost of Carbon, 2010, 2013a, 2013b, 2015a; Interagency Working Group on the Social Cost of Greenhouse Gases, 2016b). Three SC-IAMs were used: DICE (Dynamic Integrated Climate-Economy model), FUND (Framework for Uncertainty, Negotiation and Distribution model), and PAGE (Policy Analysis of the Greenhouse Effect model). Each models the relationship between CO_2 emissions and their monetized climate impact. An SC-CO_2 estimate is derived following the same causal chain for each of the SC-IAMs: a CO_2 emissions pulse is introduced in a particular year, creating a trajectory of CO_2 concentrations, temperature change, sea level rise, and climate damages.[11] The difference between this damage trajectory and the refer-

---

[9]The *Technical Support Document* was released as an appendix to rulemaking by the U.S. Department of Energy on small electric motors (Energy Conservation Program: Energy Conservation Standards for Small Electric Motors, 75 Fed. Reg. 10,874 [March 9, 2010]).

[10]There are many IAMs in use in the climate change research community for multiple purposes. Generally, IAMs vary significantly in structure, geographic resolution, computational algorithm, and application. In comparison with most other IAMs, the three used by the IWG are specialized in their focus on modeling aggregate global climate damages using highly aggregated economic and climate system representations, referred to as "reduced form IAMs": (for details, see Box 2-1 in Chapter 2). Although the three SC-IAMs were not developed solely with the purpose of estimating the SC-CO_2, they were among the very few models that produced estimates of global net economic damages from CO_2 emissions when the IWG was developing its methodology.

[11]Damages from global climate change include, but are not limited to, changes in net agricultural productivity, changes in energy use, human health effects, ocean acidification, changes in extreme weather events, and property damages from increased flood risk. Due to the long-lived nature of warming from CO_2 emissions, many of the damages from CO_2 emissions today may affect economic outcomes for the next several centuries.

ence projection in each year is discounted to the year of the CO_2 pulse using an annual discount rate.

The IWG retained most of the SC-IAMs developers' default assumptions for the parameters and functional forms used in the models. Two key exceptions are that the IWG used a single probability distribution for the equilibrium climate sensitivity (ECS)[12] parameter in all three models, as well as a common set of five future socioeconomic and emissions scenarios.[13] In addition, three constant discount rates were used to compute the present value of damages from each SC-IAM.

The IWG methodology resulted in 45 sets of estimates (three IAMs, five socioeconomic-emissions scenarios, one ECS distribution, and three discount rates) for the SC-CO_2 for a given year, with each set consisting of 10,000 estimates based on draws from the standardized ECS distribution,[14] as well as distributions of parameters treated as uncertain in two of the models. For each discount rate, the IWG combined the sets across models and socioeconomic emissions scenarios and then selected four values to be presented in regulatory impact analyses: an average value for each of three discount rates, plus a fourth value, selected as the 95th percentile of estimates based on a 3 percent discount rate. The IWG interpreted the 95th percentile as representing higher-than-expected impacts from temperature changes in the tail of the SC-CO_2 estimates: see Figure 1-1.[15]

The set of four estimates from the most recent results is shown in Table 1-1 for CO_2 impulses every 10 years from 2010 to 2050, with interim years interpolated. Percentiles and summary statistics of these estimates are presented in the IWG *Technical Support Documents*.[16]

---

[12]ECS measures the long-term response of global mean temperature to a fixed forcing, conventionally taken as an instantaneous doubling of CO_2 concentrations from their preindustrial levels (for details, see Box 4-1 in Chapter 4).

[13]The committee notes, however, that these scenarios were not fully standardized in implementation due to differences in the SC-IAMs. See Rose et al. (2014b) for details on how the IWG implemented the individual models for the estimates.

[14]The IWG selected the Roe and Baker (2007) distribution for the ECS "based on a theoretical understanding of the response of the climate system to increased greenhouse gas concentrations" and that it "better reflects the IPCC judgment that 'values substantially higher than 4.5°C still cannot be excluded'" (Interagency Working Group on the Social Cost of Carbon, 2010, pp. 13-14).

[15]The 150,000 estimates for each discount rate (2%, 3%, and 5%) are calculated by running each of the three models 10,000 times with random draws from the ECS probability distribution and other model-specific uncertain parameters, for each of the five socioeconomic emissions scenarios (three models × 10,000 runs × five socioeconomic emissions scenarios = 150,000 estimates).

[16]The full set of the most recent estimates can be found at https://obamawhitehouse. archives.gov/sites/default/files/omb/inforeg/august_2016_sc_ch4_sc_n2o_addendum_final_8_26_16.pdf [January 2017].

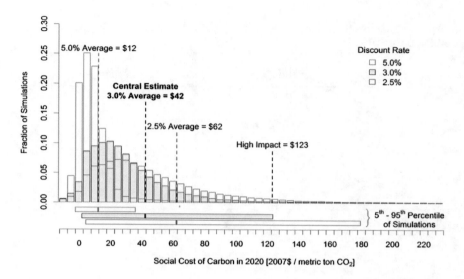

FIGURE 1-1 Frequency distributions of SC-CO$_2$ estimates for 2020 (in 2007 dollars per metric ton of CO$_2$).
NOTES: Each histogram (red, blue, green) represents model estimates, conditional on one of three discount rates, reflecting five different socioeconomic emissions scenarios, 10,000 random parameter draws, and the three SC-IAMs (see text). The frequency distributions shown represent most of the 150,000 SC-CO$_2$ estimates; however, some estimates fall outside the range shown on the horizontal axis. The *Technical Support Document* (Interagency Working Group on the Social Cost of Greenhouse Gases, 2016b) reports that 0.1 to 0.6 percent of the estimates are below the lowest bin displayed and 0.2 to 3.7 percent of the estimates are above the highest bin displayed, depending on the discount rate.
SOURCE: Interagency Working Group on the Social Cost of Greenhouse Gases (2016b, Fig. ES-1).

UPDATING ESTIMATES OF THE SOCIAL COST OF CARBON

The IWG has previously indicated its support for regular updates to the SC-CO$_2$ estimates (Working Group on the Social Cost of Carbon, 2010, p. 3): "[T]he interagency process is committed to updating these estimates as the science and economic understanding of climate change and its impacts on society improve over time." In 2013, the IWG updated the SC-CO$_2$ estimates using revised models with significant independent, model-specific updates that were made by the model developers themselves: see Table 1-2 for a summary of the model modifications.

TABLE 1-1 Social Cost of Carbon, 2010-2050 (in 2007 dollars per metric ton of CO_2)

Year	5% Average	3% Average	2.5% Average	High Impact (95th Pct at 3%)
2010	10	31	50	86
2015	11	36	56	105
2020	12	42	62	123
2025	14	46	68	138
2030	16	50	73	152
2035	18	55	78	168
2040	21	60	84	183
2045	23	64	89	197
2050	26	69	95	212

NOTE: See text for discussion.
SOURCE: Interagency Working Group on the Social Cost of Greenhouse Gases (2016b, Table ES-1).

Specifically, the IWG produced revised estimates twice in 2013 using the updated models: first in May, incorporating the revised models with IWG-specific implementation modifications,[17] and then in November, making two minor corrections to the May calculations. The IWG has continued to use these versions of the models for subsequent estimates (Interagency Working Group on the Social Cost of Carbon, 2015a; Interagency Working Group on the Social Cost of Greenhouse Gases, 2016b).

These changes resulted in an increase in the SC-CO_2 estimates reported in the 2010 *Technical Support Document* for the year 2020, which were previously reported as $7, $26, and $42 (in 2007 dollars), respectively, for the 5 percent, 3 percent, 2.5 percent discount rates and $81 for the 95th percentile at a 3 percent discount rate. The corresponding four updated SC-CO_2 estimates from the May 2013 update for 2020 were $12, $43, $65, and $129 (in 2007 dollars).

The November 2013 updates incorporated two technical corrections to the FUND modeling—correcting the potential dry land loss and the ECS distribution specification. The resulting changes to the final SC-CO_2 estimates were generally less than $1 from the May 2013 update.

The 2015 update (Interagency Working Group on the Social Cost of Carbon, 2015a, p. 21) reflected two corrections:

---

[17]Specifically, the May 2013 analysis shifted from using PAGE 2002 to PAGE09 (by Chris Hope), from DICE 2007 to DICE 2010 (by William Nordhaus), and from FUND 3.5 to FUND 3.8 (by Richard Tol and David Anthoff). See Interagency Working Group on the Social Cost of Carbon (2013a, 2013b) and Rose et al. (2014b) for descriptions of model updates and IWG modifications.

TABLE 1-2 Summary of Model Modifications Associated with the 2013 Updates of Estimates of the Social Cost of Carbon

Model	Modification
DICE	Carbon cycle parameters—weaker ocean uptake
	Sea level dynamics and valuation—explicit modeling
FUND	Space heating
	Sea level rise and land loss
	Agriculture
	Transient temperature response
	Methane—account for additional radiative forcing effects
PAGE	Sea level rise
	Revised damage function to account for saturation—modified GDP loss function
	Regional scaling factors
	Probability of discontinuity
	Adaptation
	Change in land/ocean carbon uptake
	Regional temperature change

NOTES: DICE, Dynamic Integrated Climate-Economy model; FUND, Framework for Uncertainty, Negotiation and Distribution model; GDP, gross domestic product; PAGE, Policy Analysis of the Greenhouse Effect model.
SOURCE: Adapted from Rose et al. (2014b).

First, the DICE model had been run up to 2300 rather than through 2300, as was intended, thereby leaving out the [discounted] marginal damages for the last tear of the time horizon. Second, due to an indexing error, the results from the PAGE model were in 2008 U.S. dollars rather than 2007 U.S. dollars, as was intended.

Figure 1-2 illustrates the relative values of the SC-CO_2 estimates from 2010 and 2015 for different years of CO_2 emission.

MOTIVATION FOR THE STUDY

There are significant challenges to estimating a dollar value for CO_2 emissions that reflects all of the physical and economic impacts of climate change, and the federal government made a commitment to provide regular updates to the estimates as noted above. The IWG requested this Academies study to guide future revisions of the SC-CO_2 in two important ways. First, it requested that this study provide government agencies that are part of the IWG with an assessment of the merits and challenges of a specific near-term update to the SC-CO_2 and with recommendations for enhancing the qualitative treatment or characterization of uncertainties

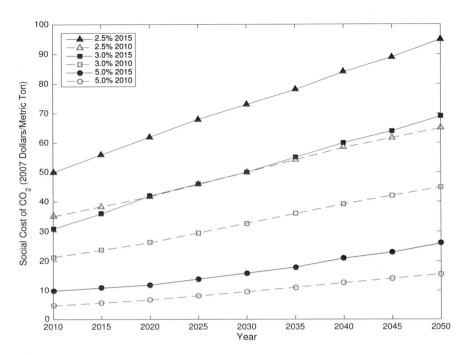

FIGURE 1-2 SC-CO_2 estimates for emissions in different years, comparing the average SC-CO_2 estimate conditional on 5, 3, and 2.5 percent discount rates. The fourth estimate is not included in this graph.
SOURCE: Data from Interagency Working Group on the Social Cost of Carbon (2010, 2015).

associated with the current SC-CO_2 estimates in their reports. The result of this request was the committee's Phase 1 report (National Academies of Science, Engineering, and Medicine, 2016). The conclusions and recommendations from the Phase 1 report are summarized in the next section.

Second, the IWG requested that the committee consider the merits and challenges of a comprehensive update of the SC-CO_2 to ensure that the estimates reflect the best available science. Specifically, it requested that the committee review the currently available science to determine its applicability for the choice of IAMs and damage functions and examine issues related to climate science modeling assumptions; socioeconomic and emissions scenarios; the presentation of uncertainty; and discounting. (The full statement of task is in Box 1-2, above.) The second phase of the study allows for broader consideration of the methodology used for estimating the SC-CO_2.

However, the statement of task was limited in its scope. Specifically, the committee was not asked to formally review or critique the current approach to estimating the SC-CO$_2$, though it did consider the current approach in making recommendations. Nor was the committee asked to consider alternatives to the use of SC-CO$_2$ estimates as a regulatory mechanism. These topics, as well as others related to the U.S. government's and other's use of SC-CO$_2$ estimates, lie outside of the scope of the committee's work and this report.

The main body of this report addresses and recommends approaches that warrant consideration in future updates of the SC-CO$_2$ estimates, as well as recommendations for research to advance the science in areas that are particularly useful for estimating the SC-CO$_2$.

SUMMARY OF STUDY'S PHASE 1 REPORT

In the Phase 1 report, the committee recommended against a near-term update to the SCC[18] estimates concluding that changing the ECS alone within the current SCC framework would not significantly improve the estimates. The committee also provided several suggestions about how to improve the communication of uncertainty in the IWG's *Technical Support Documents*. The conclusions and recommendations, grouped by the tasks they address, are in Box 1-3. The Phase 1 report also suggested that the IWG consider adopting or developing a common climate module and outlined criteria that the module should satisfy (see Chapter 4 in this report).

In August 2016 the IWG released an updated *Technical Support Document*. The IWG stated that the release responded to the committee's recommendations listed above for enhancing the presentation and improving the discussion of the uncertainty in the current estimates of SC-CO$_2$ (Interagency Working Group on the Social Cost of Greenhouse Gases, 2016b). The values for the SC-CO$_2$ estimates did not change from the 2015 release. However, the IWG provided additional material on the sources of uncertainty in the SC-IAMs in additional appendices (in response to Recommendation 2) and made the 150,000 SC-CO$_2$ values underlying each official IWG estimate available for download from the OMB website instead of by request. A new section titled "Treatment of Uncertainty" was added, together with a discussion on the types of uncertainty that are and are not included in the estimation approach (Recommendation 3).

The *Technical Support Document* also included symmetric high and low values of uncertainty in the estimates and clearly separated the values

---

[18]When referring to documents published prior to 2016, the earlier abbreviation for the social cost of carbon, SCC, is used.

BOX 1-3
Conclusions and Recommendations: Phase 1 Report

The committee's first two conclusions and its first recommendation responded to the first two of the three tasks in our statement of task: (1) whether to update the probability distribution for the ECS to reflect the recent IPCC consensus statement and (2) whether to recalibrate the distributional forms for the equilibrium climate sensitivity.

CONCLUSION 1
The equilibrium climate sensitivity (ECS) is only one parameter affecting the social cost of carbon (SCC). Each of the three SCC integrated assessment models also embodies a different representation of the climate system and its underlying uncertainties, including relationships and parameters beyond the ECS. Therefore, updating the ECS alone within the current SCC framework may not significantly improve the estimates.

CONCLUSION 2
The relationship between CO_2 emissions and global mean surface temperature can be summarized by four metrics: equilibrium climate sensitivity (ECS), transient climate response, transient climate response to emissions, and the initial pulse-adjustment timescale. ECS is less relevant than the other three metrics in characterizing the climate system response on timescales of less than a century.

As a long-term, equilibrium metric, ECS alone does not provide an adequate summary of the relationship between CO_2 emissions and global mean surface temperature for calculating the social cost of carbon (SCC). Therefore, simply updating the distribution of ECS without assessing the impact on these other metrics may not result in an improved estimate of the SCC.

RECOMMENDATION 1
The committee recommends against a near-term update to the social cost of carbon based simply on a recalibration of the probability distribution of the equilibrium climate sensitivity (ECS) to reflect the recent consensus statement in the Fifth Assessment Report of the Intergovernmental Panel on Climate Change. Consequently, the committee also recommends against a near-term change in the distributional form of the ECS.

The rest of the committee conclusions and recommendations responded to the third of our tasks, to consider enhancing the qualitative characterization of uncertainties associated with the current SCC estimates in the short-term to increase the transparency associated with using these estimates in regulatory impact analyses.

CONCLUSION 3
The Interagency Working Group on the Social Cost of Carbon (SCC) technical support document explicitly describes the factors on which the SCC is conditioned, such as the year emissions occur and the discount rate, and also makes explicit the sources of distributions for various inputs. However, it does not detail all sources of model-specific uncertainty in the social cost of carbon integrated assessment

models.

RECOMMENDATION 2

When presenting the social cost of carbon (SCC) estimates, the Interagency Working Group (IWG) on the SCC should continue to make explicit the sources of uncertainty. The IWG should also enhance its efforts to describe uncertainty by adding an appendix to the technical support document that describes the uncertain parameters in the Climate Framework for Uncertainty, Negotiation and Distribution and Policy Analysis of the Greenhouse Effect models.

CONCLUSION 4

Multiple runs from three models provide a frequency distribution of the social cost of carbon (SCC) estimates based on five socioeconomic-emissions scenarios, three discount rates, draws from the equilibrium climate sensitivity distribution, and other model-specific uncertain parameters. This set of estimates does not yield a probability distribution that fully characterizes uncertainty about the SCC.

RECOMMENDATION 3

The Interagency Working Group on the Social Cost of Carbon (IWG) should expand its discussion of the sources of uncertainty in inputs used to estimate the social cost of carbon (SCC), when presenting uncertainty in the SCC estimates. The IWG should include a section entitled "Treatment of Uncertainty" in each technical support document updating the SCC. This section should discuss various types of uncertainty and how they were handled in estimating the SCC, as well as sources of uncertainty that are not captured in current SCC estimates.

CONCLUSION 5

It is important to continue to separate the impact of the discount rate on the social cost of carbon from the impact of other sources of variability. A balanced presentation of uncertainty includes both low and high values conditioned on each discount rate.

RECOMMENDATION 4

The executive summary of each technical support document should provide guidance concerning interpretation of reported social cost of carbon (SCC) estimates for cost-benefit analysis. In particular, the guidance should indicate that SCC estimates conditioned on a particular discount rate should be combined with other cost and benefit estimates conditioned on consistent discount rates, when they are used together in a particular analysis.

The guidance should also indicate that when uncertainty ranges are presented in an analysis, those ranges should include uncertainty derived from the frequency distribution of SCC estimates. To facilitate such inclusion, the executive summary of the technical support document should present symmetric high and low values from the frequency distribution of SCC estimates with equal prominence, conditional on each assumed discount rate.

NOTE: The committee's Phase 1 report used the then-current acronym for the social cost of carbon, SCC.

by the discount rate, as shown by the bars below the graph in Figure 1-1 (above) (Recommendation 4). The IWG continues to emphasize the non-symmetric uncertainty in the estimates by including the 95th percentile values in the executive summary table (see Table 1-1, above) despite the committee's Phase 1 recommendation to present symmetric high and low values from the frequency distribution of SCC estimates with equal prominence, conditional on each assumed discount rate (Recommendation 4; Box 1-3, above). Agencies continue to be directed to use these estimates, but are able to conduct sensitivity analysis if an agency determines it appropriate. Agencies are referred to OMB Circular A-4 for best practices in conducting uncertainty analysis in RIAs.

The IWG also released in August 2016 an addendum to the updated *Technical Support Document* on estimating the social costs of methane (CH_4) and nitrous oxide (N_2O) (Interagency Working Group on the Social Cost of Greenhouse Gases, 2016a), noting that the framework for the non-CO_2 estimates is the same as that used for SC-CO_2 estimation. This report does not review or assess these new estimates for CH_4 and N_2O.

STRATEGY TO ADDRESS THE STUDY CHARGE

This study was carried out by a committee of experts appointed by the president of the National Academy of Sciences. The committee consisted of 13 members, working with a technical consultant and study director. Committee expertise spans the issues relevant to the study task: integrated assessment modeling, statistical modeling, climate science, climate impacts, environmental economics, energy economics, decision science, public policy, and regulation. In selecting the committee, care was taken to ensure that the membership possesses the necessary balance between research and practice by including academic scientists and other experts. Committee members were chosen to have the relevant disciplinary expertise and to ensure there are no current connections that might constitute a conflict of interest with the Department of Energy, the Environmental Protection Agency, or other regulatory agency members of the IWG. Biographical sketches of the committee members and staff are provided in Appendix A.

To address the Phase 2 task, the committee held three open meetings to receive information from federal agency staff to understand its study charge and to gather information to explore its charge (see Appendix B). Closed sessions were held to refine and finalize the committee's conclusions and recommendations. The project included two focused studies by outside experts to support the committee's analyses: a study on global growth projections as applied to the SC-CO_2 estimation problem (see

Chapter 3 and Appendix D) and a literature review of climate damages and impacts (the results of which are used in Chapter 5).

The report is organized in seven chapters, with the committee's conclusions and recommendations included in the relevant chapters, and several appendices. Chapter 2 provides a high-level response to the statement of task and an overview of the framework the committee used. Chapters 3-6 provide specific details and recommendations on the implementation of both near-term and longer-term updates. Chapter 3 is focused on updates to socioeconomic and emissions projections; Chapter 4 considers updates to modeling of the Earth system, including temperature change, sea level rise, and ocean acidification; Chapter 5 explores updates to climate impacts and damage estimates; and Chapter 6 presents an updated approach to discounting future damages. Chapter 7 highlights research priorities in key areas that are needed to improve future updates to the SC-CO$_2$ estimates by summarizing research conclusions found throughout the report.

The five substantive appendices provide further technical detail on specific subjects: expert elicitation (Appendix C), projections of global economic growth (Appendix D), calculation of ocean acidification (Appendix E), comparison of the climate components of the SC-IAMs with a simple Earth system model (Appendix F), and model-specific suggestions for near-term improvements to current SC-IAMs damage components (Appendix G).

2

Framework for Estimating
the Social Cost of Carbon

This chapter provides an overview of the steps involved in estimating the social cost of carbon and the committee's recommendations for how they should be organized in future updates. The committee discusses how uncertainty might be characterized in such a framework and the level of geographic, sectoral, and temporal detail involved. The frequency of updates to SC-CO_2 estimates and how the process of updating SC-CO_2 estimates might be structured is also discussed.

STRUCTURE OF THE ESTIMATION PROCESS

Estimating the SC-CO_2 involves four steps: (1) projecting future global and regional population, output, and emissions; (2) calculating the effect of emissions on temperature, sea level, and other climate variables; (3) estimating (explicitly or implicitly) the physical impacts of climate and, to the extent possible, monetizing those impacts on human welfare (i.e., estimating net climate damages); and (4) discounting monetary damages to the year of emission.

The committee structured its work, conclusions, and recommendations around these four parts of a framework—socioeconomic factors and emissions, climate, impacts and damages, and discounting—which are characterized as modules. Each of these modules is comprised of data, conceptual formulations and theory, computer models and other analysis frameworks. And, to some extent, each is supported by its own specialized disciplinary expertise. Estimation of the SC-CO_2 involves the integration of these four modules, taking account when possible the inter-dependencies among them.

Studies supporting SC-CO$_2$ estimation have used integrated assessment models (IAMs) that incorporate some or all of the four components in a single model: Box 2-1 details some key terminology related to model-

---

BOX 2-1
Modeling Terminology in This Report

Integrated assessment models. IAMS are computational models of global climate change that include representation of the global economy and greenhouse gas emissions, the response of the climate system to human intervention, and impacts of climate change on the human system. IAMs fall into two general categories: detailed-structure IAMs and reduced-form IAMs.

Detailed-structure IAMs. These models have a regional and sectoral economic structure that were originally developed to study the effects of technology and policy on greenhouse gas emissions (e.g., Edmonds and Reilly, 1983). Increasingly, detailed-structure IAMs have begun to include some elements of impacts and adaptation (e.g., Reilly et al., 2012a; Calvin et al., 2013). They are used to assess climate change risk, study detailed climate mitigation policy proposals, and investigate climate impacts by sector and region. They also are used to study the interactions between different climate change impact sectors such as agriculture, water, energy and land, and to study the feedbacks from these sectors to the climate system. Since the climate change impact sectors in these models are represented by physical system representations that are spatially and structurally explicit, the interactions between those impact areas, the socioeconomic system, and the climate system can each be tracked at a variety of geographical scales. Although none of these models has yet been used to comprehensively evaluate global physical and socioeconomic impacts, or to sum all potential climate change damages, they have been used to study a number of potentially important interactions and feedbacks (e.g., Wise et al., 2009; Reilly et al., 2012a) and to study these systems more comprehensively at the regional level (e.g., Kraucunas et al., 2015). Examples of detailed-structure IAMs include the Global Change Assessment Model (GCAM) (Joint Global Change Research Institute, 2015), the Integrated Global System Modeling Framework of the Massachusetts Institute of Technology (IGSM) (Reilly et al., 2012a), anthropogenic emission prediction and policy analysis (EPPA) (Chen et al., 2016), the Asian-Pacific Integrated Model (AIM) (Matsuoka et al., 1995), An Integrated Model to Assess the Greenhouse Effect (IMAGE) (Rotmans, 1990), the World Induced Technical Change Hybrid Model (WITCH) (Bosetti et al., 2006), and the Model for Energy Supply Strategy Alternatives and their General Environmental Impact (MESSAGE) (Agnew et al., 1978).

Reduced-form IAMs. These highly aggregated models include representation of global climate damages. They were originally developed (e.g., Nordhaus, 1994b)

ing used in this report. The IWG used three reduced-form IAMs—DICE (Dynamic Integrated Climate-Economy model), FUND (Framework for Uncertainty, Negotiation and Distribution model), and PAGE (Policy Analysis of the Greenhouse Effect model)—to compute the SC-CO$_2$, pooling the final SC-CO$_2$ estimates from each model at the end of the analysis. The essence of the committee's approach is to unbundle the four steps of

to study optimal global CO$_2$ emissions trajectories and carbon prices that maximize global welfare. A second application is to compute the costs and benefits of policies that seek to achieve climate objectives other than welfare maximization. The DICE (Dynamic Integrated Climate-Economy Model), FUND (Framework for Uncertainty, Negotiation and Distribution), PAGE (Policy Analysis of the Greenhouse Effect), and ENVISAGE (ENVironmental Impact and Sustainability Applied General Equilibrium) (Roson and van der Mensbrugghe, 2012) models fall into this category. In contrast to detailed-structure IAMs, they attempt to represent comprehensively the impacts of climate change on human welfare, but they do not attempt to provide a detailed structural model of the global economy.

SC-IAMs. Since there are many IAMs in use in the climate change research community, for multiple purposes, this report refers to the three reduced-form IAMs used by the IWG as SC-IAMs. Generally, IAMs vary significantly in structure, geographic resolution, computational algorithms, and applications. In comparison with most other IAMs, the three SC-IAMs used by the IWG—DICE, FUND, and PAGE—are specialized in their focus on modeling aggregate global climate damages using highly aggregated economic and climate system representations. Although the three SC-IAMs were not developed solely to estimate the SC-CO$_2$, they are among the very few models that can be used to estimate global net economic damages from CO$_2$ emissions.

SC-CO$_2$ estimation. The committee uses "methodology" when referring to the IWG's current estimation process; "framework" when referring to this committee's proposed approach; and "process" as a generic term to describe a set of analyses.

Parts of SC-CO$_2$ estimation. Several terms are used throughout the report to refer to specific aspects of SC-CO$_2$ estimation. "Model" refers to an existing modeling system, including the three SC-IAMs used in the current SC-CO$_2$ methodology: DICE, FUND, and PAGE. "Component" describes the parts of the existing SC-IAMs. Each model in the current SC-CO$_2$ methodology contains four components: socioeconomic, climate, damages, and discounting. "Modules" describes the parts of the committee's proposed framework, and it has the same four elements: socioeconomic, climate, damages, and discounting. Words such as "formulation," "element," and "function" describe specific relationships within modules, components, or models (e.g., the agriculture sector damages formulation in an SC-IAM).

the analysis. Rather than averaging the results from three separate SC-IAMs, the committee suggests a single framework with modules designed to capture uncertainty at each step. This report focuses on how each module of the analysis could be constructed.

Figure 2-1 illustrates the parts of the SC-CO$_2$ estimation process in terms of the committee's recommended modular framework, showing how information is exchanged among the modules, leading ultimately to the SC-CO$_2$ estimate. The flow of intermediate results in the framework is shown in solid lines. The dashed lines introduce additional possible interactions among components of the estimation that are discussed below.

As Figure 2-1 shows, a socioeconomic module (detailed in Chapter 3) generates projections of greenhouse gas emissions for input to the climate module, as well as estimates of future population and gross domestic product (GDP) that are direct inputs to the damages module and the discounting module (Box 2-2 details the key economic and related terms used in this report). The projected emissions paths serve as a baseline to which an emission pulse is added in order to represent the incremental impact of an additional ton of CO$_2$ released in a particular year.

Based on projected emissions, a climate module (detailed in Chapter 4) generates estimates of greenhouse gas concentrations, temperature change, sea level rise, and other needed climate variables. Along with population and GDP, these climate results are then inputs into a damages module (detailed in Chapter 5) that calculates the physical impacts of climate variables on outcomes that affect human welfare and, when possible, monetizes on a year-by-year basis the net damages caused by the climate change due to CO$_2$ emissions. The grey dashed outline around the damages module in Figure 2-1 indicates that regional or sectoral socioeconomic and climate data will likely be necessary either as direct inputs to impact functions or for their calibration. The figure also shows that non-monetized impacts may also be included in representations of the cost of CO$_2$ emissions, albeit in physical rather than monetary terms.

The purpose of a discounting module (detailed in Chapter 6) is to integrate the future stream of monetized damage estimates into a single present value for each state of the world generated by the earlier steps of the SC-CO$_2$ estimation process. The committee suggests an approach yielding three discount rate scenarios, with each scenario having a distribution of SC-CO$_2$ values that is determined by all the other sources of uncertainty incorporated in the SC-CO$_2$ estimation process. To date, the IWG has focused on scenarios with fixed discount rates of 2.5 percent, 3 percent, and 5 percent. The dotted line in Figure 2-1 shows GDP and population as recommended future inputs into the discounting module to capture the relationship between the year-to-year discount rate and growth in per capita GDP.

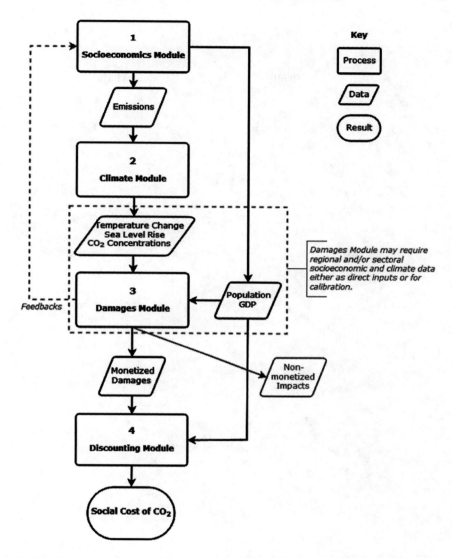

FIGURE 2-1 An integrated, modular framework for estimating the social cost of carbon (SC-CO$_2$).

NOTES: The figure shows the different modules that are involved in the computation of the SC-CO$_2$ and the possible linkages among them. An integrated, modular framework focuses on developing each module of the analysis using the criteria outlined in this report and combining them to estimate the SC-CO$_2$. See text for discussion.

BOX 2-2
Economic and Related Terms Used in the Report

Gross domestic product (GDP) and related terms. GDP represents the value of all goods and services produced by a country and explicitly or implicitly sold in markets. The report uses GDP when referring to all countries (also known as gross world product [GWP]). GDP divided by population is referred to as per capita GDP. A related concept, gross national income (GNI) represents the market income earned by all citizens of a country. When summed across all countries, GNI is equal to GDP. GDP and GNI as currently measured exclude some components of production and income: household production (e.g., cooking, cleaning, and child-care for one's own family) is generally excluded from GDP and GNI, and production that occurs outside of formal markets may also be excluded.

Consumption. Consumption refers to the value of all goods and services consumed by households. Some of these may be purchased in markets, and thus constitute part of GDP, while others (e.g., good health, ecosystem services) are not generally traded in markets.

Capital stock. Capital stock refers to productive assets, which can be physical (e.g., factories, bridges), human (e.g., the stock of knowledge embodied in a population), or natural (e.g., land, wetlands, animals).

Investment. Investment refers to expenditures to increase the capital stock.

Impacts of climate change. Impacts include the physical effects of temperature, sea level rise and other climate variables on production, consumption, investment, and capital stocks. They also include physical impacts that do not directly affect markets, such as impacts on biodiversity.

Damages from climate change. Damages are the monetized value of the net impacts associated with climate change. Conceptually, the economic measure of climate damages is what consumers would be willing to pay to avoid such changes to the climate. In practice, because of data limitations, damages are often measured in terms of their effect on GDP.

AN INTEGRATED, MODULAR FRAMEWORK

Current estimates of the SC-CO$_2$ are obtained by averaging estimates of monetized damages produced by the three SC-IAMs, each of which contains its own climate component and set of damage functions. Although a common set of socioeconomic scenarios and a common distribution of equilibrium climate sensitivity[1] (ECS) is applied to each SC-IAM, their climate and damage components differ significantly. The final distribution over the SC-CO_2 estimates is based on the average and range of the different components of these structurally distinct models (see Figure 1-1, in Chapter 1).

A previous study has documented the differences in the assumptions and functional forms embedded in the climate components of the three SC-IAMs (Rose et al., 2014b). Of particular concern to the committee is that, even under a common value of the ECS, significant differences in climate modeling structure and climatic response underlie the estimates from the three SC-IAMs that are being averaged. Such differences would be informative if they systematically represented structural uncertainty in the climate system—that is, uncertainty about which is the correct modeling structure to use—but in practice, and as discussed in more detail in Chapter 4, the differences arise instead from uncoordinated modeling choices of the individual model developers.

Because of these differences, in its Phase 1 report (National Academies of Science, Engineering, and Medicine, 2016) the committee suggested that the IWG undertake efforts to adopt or develop a common climate module with three characteristics:

(1) It is consistent with the best available scientific understanding of the relationship between emissions and temperature change, its pattern over time, and its uncertainty.
(2) It strives for simplicity and transparency so that the central tendency and range of uncertainty in its behavior are readily understood, are reproducible, and are amenable to continuous improvement over time through the incorporation of evolving scientific evidence.
(3) It considers the possible implications of the choice of a common climate module for the assessment of impacts of other, non-CO_2 greenhouse gases.

A similar argument for a common module can be made for the socioeconomic and impact/damage components of the analysis, in effect indicating an approach to the IWG's task that places heavy emphasis on improving the scientific and information basis of each of the four main components of the analysis. The committee contends that this modular approach is superior to averaging the SC-CO_2 estimates from separate IAMs that may depend on inconsistent assumptions or on assumptions that, when averaged, do not yield an overall distribution of uncertainty that is consistent with the best available evidence.

---

[1]ECS measures the long-term response of global mean temperature to a fixed forcing, conventionally taken as an instantaneous doubling of CO_2 concentrations from their preindustrial levels. See Chapter 4 for further discussion.

Such a modular estimation framework can help ensure consistency of key assumptions and can aid in the rigorous and transparent characterization of uncertainty at each stage of the estimation process. It can also provide a means for transparently identifying the inputs, outputs, and linkages among the various stages of the SC-CO$_2$ estimation process. This modularity can thereby enable expert groups in the broad scientific community to evaluate aspects of the process that are within their disciplinary expertise, while ensuring that these elements are coherently integrated.

The main risk in a focus on individual modules is a failure to identify and take proper account of feedbacks and other interactions among components of the human-climate system that cut across these modular boundaries. This concern is addressed below, as well as potential future research activities that could be undertaken to address it.

CONCLUSION 2-1 For at least some steps in the SC-CO$_2$ estimation framework, using a common module—rather than averaging the results from multiple models—can improve transparency and consistency of key assumptions with the peer-reviewed science and can improve control over the uncertainty representation, including structural uncertainty. This rationale underlies the Interagency Working Group's use of the same socioeconomic scenarios, discount rates, and distribution for climate sensitivity across IAMs, as well as the committee's suggestion in its Phase 1 report that the IWG develop or adopt a common climate module.

CONCLUSION 2-2 An integrated modular framework for SC-CO$_2$ estimation can provide a transparent identification of the inputs, outputs, uncertainties, and linkages among the different steps of the SC-CO$_2$ estimation process. This framework can also provide a mechanism for incorporating of new scientific evidence and for facilitating regular improvement of the framework modules and resulting estimates by engaging experts across the varied disciplines that are relevant to each module.

RECOMMENDATION 2-1 The Interagency Working Group should support the creation of an integrated modular SC-CO$_2$ framework that provides a transparent articulation of the inputs, outputs, uncertainties, and linkages among the different steps of SC-CO$_2$ estimation. For some modules within this framework, the best course of action may be for the government to develop a new module, while for other modules the best

course of action may be to adapt one or more existing models developed by the scientific community.

The committee recognizes that models developed in academic research may require substantial modification before being appropriate for use in estimating the SC-CO$_2$, as the purpose of that research is not to generate estimates for use in regulatory impact analysis. The committee leaves to the discretion of the IWG the best way to assemble results from the scientific literature into a modular framework for estimating the SC-CO$_2$. This may involve issuing contracts to researchers outside of the U.S. government and/or choosing to have the analysis performed within the government.

Subsequent chapters outline criteria that are specific to each module: below is a general set of standards that apply to all analytical efforts to estimate the SC-CO$_2$.[2]

RECOMMENDATION 2-2 The Interagency Working Group should use three criteria to evaluate the overall integrated SC-CO$_2$ framework and the modules to be used in that framework: scientific basis, uncertainty characterization, and transparency.

- **Scientific basis: Modules, their components, their interactions, and their implementation should be consistent with the state of scientific knowledge as reflected in the body of current, peer-reviewed literature.**
- **Uncertainty characterization: Key uncertainties and sensitivities, including functional form, parameter assumptions, and data inputs, should be adequately identified and represented in each module. Uncertainties that cannot be or have not been quantified should be identified.**
- **Transparency: Documentation and presentation of results should be adequate for the scientific community to understand and assess the modules. Documentation should explain and justify design choices, including such features as model structure, functional form, parameter assumptions, and data inputs, as well as how multiple lines of evidence are combined. The extent to which features are evidence-**

---

[2]The committee notes that the criteria listed in Recommendation 2-2 reinforce and are consistent with U.S. Office of Management and Budget (OMB) Circular A-4 guidance for regulatory impact analysis, which includes general guidelines for "Transparency and Reproducibility of Results" and the "Treatment of Uncertainty."

based or judgment-based should be explicit. Model code should be available for review, use, and modification by researchers.

FEEDBACKS AND INTERACTIONS

Over time, successful implementation of such a framework will require attention to the interactions among the modules, and necessitate adaptation of the overall structure to take advantage of ongoing research on the human-environment-climate system. One example of such interaction is suggested in Figure 2-1 (above): a dashed line indicates that climate damages, evaluated in the damages module, could feed back onto greenhouse gas emissions, as represented in a socioeconomic module. If the output of the damages module shows a reduction in GDP, for example, that reduction may affect the projected GDP in subsequent years and thus the projected emissions from the socioeconomic module. Similarly, the temperature- and CO_2-driven impacts on crop yields projected by the damages module may affect agricultural productivity in the socioeconomic module, and the warming-driven impacts on air-conditioner adoption and use could affect energy use and thus CO_2 emissions projected by the socioeconomics module. In the current SC-CO_2 estimation, the emissions projections are exogenous in all three SC-IAMs and GDP projections are exogenous in two of the three SC-IAMs. In the current framework, there is therefore little feedback from climate or damages to emissions and GDP.

Ongoing research on climate impacts/damages, integrated assessment, and Earth system modeling is revealing interactions among the components of SC-CO_2 estimation that go beyond the above examples of feedback from climate impacts/damages to socioeconomic projections, suggesting additional future changes in implementation of the four-module framework recommended in this report. For example, it is now recognized that some of the most severe impacts of climate change on particular regions in specific sectors result from interactions between them and impacts in other regions or sectors (Oppenheimer et al., 2014). These interactions can occur at the physical level: for example, if climate change causes a particular region to become hotter and dryer, it might increase the competition for limited water between agricultural, power plant cooling, and household and commercial uses (see Taheripour et al., 2013). These shortages could then make both food and energy more expensive in some regions, which could have general equilibrium effects on other economic sectors in that region or elsewhere (Baldos et al., 2014). Indeed, increases in temperature and decreases in precipitation are already occurring in a number of major growing regions, leading to the need for more irrigation, which is largely being met by depleting ground water resources (Grogan et al., 2015). As ground water aquifers are drained, water becomes even more scarce, which may reduce agricultural production in some very vulnerable low-income regions (Zaveri et al., 2016). Other interactions may mitigate impacts, including, for example, the reduction of the health impacts of heat stress by increased air conditioning.

Such interactions are being explored in detailed-structure IAMs (Reilly et al., 2012a). One prominent example is the interaction between greenhouse gas mitigation and urban and regional air pollution policies (Reilly et al., 2007; Chuwah et al., 2013; Nam et al., 2014). Another is the study of competition for water for both agriculture and power plant cooling that will occur in hotter and dryer climates, as well as of the impacts on water and land (and the resulting land emissions) of global policies that rely on massive increases in biofuels (Reilly et al., 2012b; Daioglu et al., 2014; Rose et al., 2014a).

Although the literature on these feedbacks and interactions is advancing, many of the relevant studies consider these phenomena one at a time, perhaps missing interactions among them. Some yield results only in physical terms and do not proceed to economic measures, and the studies to date frequently consider only one country or region and do not provide a basis for extension to the world as a whole. Thus, opportunities for incorporating the relevant feedbacks and interactions in a modular approach depend on the state of scientific knowledge as it will emerge from ongoing research, as well as on details of the damage functions and the nature of the modeling frameworks used for the socioeconomics projections.

Even at the simplest level of interaction among modules, shown in Figure 2-1, there will be need for careful attention to the flow of information among them. And care will be required during the development phase to maintain consistency among modules. For example, components will require consistency or appropriate conversion across units of measurement, time steps, uncertainty representation, and regional and sectoral specification. Additionally, the modules may differ in their choice of software development systems. This task will grow more challenging with consideration of additional feedbacks and other interactions that will require stronger and tighter coupling of the modules. It will thus be desirable to choose an integration platform that can accommodate change in internal module structure and interactions; it may even be desirable to integrate the components into a single computational framework.

CONCLUSION 2-3 Research to identify and explore the magnitude of various interactions and feedbacks within the human-climate system, which are relationships not currently well

represented in the SC-CO_2 estimation framework, will be an important input to longer-term enhancements in the SC-CO_2 estimation framework. Areas of research that are likely to yield particular benefits include:

- Exploration of methods for representing feedbacks among systems and interactions within them, such as:
  - feedbacks between climate, physical impacts, economic damages, and socioeconomic projections, and
  - interactions between types of impacts or economic damages within and across regions of the world.
- Assessment of the relative importance of specific feedbacks and interactions in the estimation of the SC-CO_2, perhaps using an existing detailed structure model of the world economy.
- Assessment of existing analyses that integrate socioeconomic, climate, and damage components to assess their suitability for use in estimating the SC-CO_2, particularly with respect to feedbacks and interactions, while recognizing the computational requirements for such analyses.

RECOMMENDATION 2-3 The Interagency Working Group should continue to monitor research that identifies and explores the magnitude of various interactions and feedbacks in the human-climate system including those not represented in implementation of the proposed modular SC-CO_2 estimation framework. The IWG should include interactions and feedbacks among the modules of the SC-CO_2 framework if they are found to significantly affect SC-CO_2 estimates.

GEOGRAPHIC, SECTORAL, AND TEMPORAL DETAIL

Implementation of a modular approach will entail decisions about the level of regional and sectoral detail in each module and about the time horizon over which estimates are computed. These issues are discussed below, with specific focus on the United States.

Level of Geographic and Sectoral Detail

As the dashed box in Figure 2-1 indicates, estimation and implementation of damage functions may require climate and/or socioeconomic inputs at a regional and/or sectoral level. The level of regional and sector disaggregation necessary will be dictated by the level of disaggregation of damages. The damage module may specify separate damage functions for different impact sectors (e.g., agriculture, sea level rise, electricity generation) and different regions (e.g., United States, India, China, sub-Saharan Africa). Disaggregation may also be necessary as an intermediary step toward calibration of an aggregate damage function. Chapter 3 discusses approaches for providing disaggregated population and GDP by region and sector, and Chapter 4 discusses methods for regional downscaling of temperature and sea level rise.

In the near term, probabilistic spatially disaggregated projections of population, GDP, temperature and sea level rise are particularly challenging. However, deterministic disaggregation (i.e., point estimates of variables) could be used as direct inputs to regional and sectoral damage functions or to calibrate global damage functions based on detailed regional and sectoral damage characterizations.

In the longer run, it may be possible to use models with regional and sectoral detail as the basis of socioeconomic projections and to provide probability distributions of disaggregated values for population and GDP. Similarly, it may be possible in the longer term to obtain probability distributions defined over spatially disaggregated climate variables.

U.S. Damages

Because CO_2 emissions have global impacts regardless of the country from which they originate, both research and IWG efforts to estimate the SC-CO_2 have focused on total global damages, rather than the damages to any individual country. In 2010 the Interagency Working Group on the Social Cost of Carbon (2010, p. 10) stated that "because of the distinctive nature of the climate change problem, we center our current attention on a global measure of SCC [SC-CO_2]." At the same time, the IWG recognized that this approach "represents a departure from past practices, which tended to put greater emphasis on a domestic measure of SCC (limited to impacts of climate change experienced within U.S. borders)." Nonetheless, the IWG asserted its flexibility, noting that (p. 10):

> [A]s a matter of law, consideration of both global and domestic values is generally permissible; the relevant statutory provisions are usually ambiguous and allow selection of either measure. . .

> Under current OMB guidance contained in Circular A-4, analysis of economically significant proposed and final regulations from the domestic perspective is required, while analysis from the international perspective is optional.

In its updates to the SC-CO_2, the IWG has consistently supported a focus on global values, as reflected in many *Technical Support Document* updates (Interagency Working Group on the Social Cost of Carbon, 2010, pp. 10-11; 2013a, p. 14; 2013b, p. 14; 2015, p. 14):[3]

> [T]he climate change problem is highly unusual in at least two respects. First, it involves a global externality: emissions of most greenhouse gases contribute to damages around the world even when they are emitted in the United States. Consequently, to address the global nature of the problem, the SCC must incorporate the full (global) damages caused by GHG [greenhouse gas] emissions. Second, climate change presents a problem that the United States alone cannot solve. Even if the United States were to reduce its greenhouse gas emissions to zero, that step would be far from enough to avoid substantial climate change. Other countries would also need to take action to reduce emissions if significant changes in the global climate are to be avoided. Emphasizing the need for a global solution to a global problem, the United States has been actively involved in seeking international agreements to reduce emissions and in encouraging other nations, including major emerging major economies, to take significant steps to reduce emissions. When these considerations are taken as a whole, the interagency group concluded that a global measure of the benefits from reducing U.S. emissions is preferable.

Despite this consistent focus, the IWG did explore the basis for estimating the SC-CO$_2$ for the United States (Interagency Working Group on the Social Cost of Carbon, 2010, p. 11): "[A]s an empirical matter, the development of a domestic SCC is greatly complicated by the relatively few region- or country-specific estimates of the SCC in the literature." Using only the FUND model (which has regional disaggregation to its damage functions), the IWG noted an average U.S. benefit of about 7-10 percent of the global benefit across the scenarios it analyzed. Alternatively, the IWG found that "if the fraction of GDP lost due to climate change is assumed to be similar across countries, the domestic benefit would be proportional to the U.S. share of global GDP, which is currently about 23 percent." On this basis, the IWG "determined that a range of values from 7 to 23 percent should be used to adjust the global SCC to calculate domestic effects," recognizing that "these values are approximate, provisional, and highly speculative." Nonetheless, as described in the IWG's 2015 *Response to Comments* (Interagency Working Group on the Social Cost of Carbon, 2015b, p. 31), some commenters have asserted that domestic damage estimates have received inadequate attention.

Correctly calculating the portion of the SC-CO$_2$ that directly affects the United States involves more than examining the direct impacts of

---

[3]The 2016 update of the *Technical Support Document* uses similar language on p.17 (Interagency Working Group on the Social Cost of Greenhouse Gases, 2016b) but also cites the worldwide commitment by many countries to reducing greenhouse gases in the signing of the Paris Agreement on April 22, 2016.

climate that occur within the country's physical borders, which is what the 7-23 percent range is intended to capture. Climate damages to the United States cannot be accurately characterized without accounting for consequences outside U.S. borders. As the IWG noted (Interagency Working Group on the Social Cost of Carbon, 2010), climate change in other regions of the world could affect the United States through such pathways as global migration, economic destabilization, and political destabilization. In addition, the United States could be affected by changes in economic conditions of its trading partners: lower economic growth in other regions could reduce demand for U.S. exports, and lower productivity could increase the prices of U.S. imports. The current SC-IAMs do not fully account for these types of interactions among the United States and other nations or world regions in a manner that allows for the estimation of comprehensive impacts for the United States.

In addition, the United States may choose to use a global SC-CO_2 in order to leverage reciprocal measures by other countries (Kopp and Mignone, 2013; Howard and Schwartz, 2016). U.S. emissions impose most of their damage beyond U.S. borders, and climate damages to U.S. citizens will largely depend on emissions and mitigation choices by other countries. Climate damages and mitigation benefits to each country are thus determined by the global effort, and the potential to leverage foreign mitigation supports a domestic SC-CO_2 estimate augmented by the expected foreign leverage (Pizer et al., 2014). Considering all these factors, there are reasons to consider a global SC-CO_2 and what constitutes domestic impact in the case of a global pollutant.

> **CONCLUSION 2-4 Estimation of the net damages per ton of CO_2 emissions to the United States alone, beyond the approximations done by the IWG, is feasible in principle; however, it is limited in practice by the existing SC-IAM methodologies, which focus primarily on global estimates and do not model all relevant interactions among regions. It is important to consider what constitutes a domestic impact in the case of a global pollutant that could have international implications that impact the United States. More thoroughly estimating a domestic SC-CO_2 would therefore need to consider the potential implications of climate impacts on, and actions by, other countries, which also have impacts on the United States.**

Time Horizon

In concept, the SC-CO_2 assesses the total discounted damage to social welfare caused by an emission of CO_2 occurring in a particular year, which results in damages that can persist several centuries into the future. Thus, the question arises of what time horizon should be used for the analysis. In the context of the socioeconomic, damage, and discounting assumptions, the time horizon needs to be long enough to capture the vast majority of the present value of damages.[4] The length of this horizon depends on the rate at which undiscounted damages grow over time and on the rate at which they are discounted. Longer time horizons allow for representation and evaluation of longer-run geophysical system dynamics, such as sea level change and the carbon cycle; however, they involve greater speculation and uncertainty about socioeconomic conditions and emissions. It will be informative, for analytic transparency and decision making, for the IWG to report the share of the SC-CO$_2$ accruing over different time horizons. Such reporting would provide a sense of the relative importance of very long-term impacts to the overall estimate.

TREATMENT OF UNCERTAINTY

The inputs and the outputs of each module of the SC-CO$_2$ analysis have inherent uncertainties, as do the structures and parameters of the modules themselves. The future growth rates of population and output are uncertain. The emissions associated with any future GDP path depend on policies to control greenhouse gas emissions and on the evolution of energy technologies, energy markets, and land-use patterns—all of which are uncertain. Although the basic physics of the climate system are well established, the parameters linking emissions to mean global temperature and other climate variables are not known with certainty. Similarly, there is uncertainty in the translation of physical climate changes into impacts and damages. Given these numerous uncertainties inherent in SC-CO$_2$ estimation, the IWG necessarily needs to focus its analytical attention on incorporating the most important sources of uncertainty, rather than seeking to incorporate all possible sources of uncertainty.

Both parametric uncertainty and structural uncertainty are in each of the modules that comprise the SC-CO$_2$ framework. Parametric uncertainty is uncertainty about the value of various parameters in a model (or models) used in a module. A physical climate example of parametric uncertainty is uncertainty in the strength of known feedbacks that amplify or dampen the sensitivity of the global mean temperature to climate forcing. Structural uncertainty is uncertainty about what model constitutes the best framework for understanding what one wishes to project. A physical climate example of structural uncertainty is the possible presence

---

[4]"Vast majority" is a deliberately vague term that signals much more than 50 percent, but not 100 percent.

of unknown feedbacks, which may be hinted at by the geological record of past climate responses to climate forcing similar to the magnitude of recent anthropogenic forcing. Another example of structural uncertainty arises with respect to climate damages because of unknown damage pathways associated with abrupt climate change.

Approaches to Decision Making under Uncertainty

The standard approach to benefit-cost evaluation under uncertainty is expected value analysis of the consequences for social welfare, with additional consideration of the distribution of consequences around the expected value. As explained in Chapter 1, this is the regulatory approach that underlies regulatory impact analysis under Executive Order 12866, for which the SC-CO$_2$ was developed. Under this approach, one evaluates each regulation by the resulting change in expected welfare, that is, by the effect of that regulation on social welfare in each possible state of the world weighted by the probability of that state. The results of this approach depend on the probabilities associated with each state of the world.

Other approaches make use of multiple probability distributions over states of the world. One such approach is max-min expected utility, which recommends the policy for which the minimum expected utility, calculated by using the alternative probability distributions, is as large as possible. Another approach requires one to assign subjective weights to each of the alternative probability distributions and recommends the policy for which the weighted average expected utility is maximized. An advantage of these approaches is that they can incorporate multiple probability distributions that are consistent with available information without the need to select a unique probability distribution, as occurs with expected utility. These other approaches can also incorporate ambiguity aversion, for example, when a decision maker prefers a policy for which there is greater confidence about the probabilities. However, these approaches can be sensitive to the exact set of probability distributions that are considered, as well as to the assignment of weights to these distributions.

As discussed in Chapter 1, the committee interpreted its charge as focusing specifically on the SC-CO$_2$ for its use in regulatory impact analysis. The committee therefore developed its conclusions and recommendations to be consistent with an overall analytical approach based on the computation of expected net present value, taking into account that the SC-CO$_2$ is one of a large number of additional variables that enter into a typical regulatory impact analysis.

The committee recognized that the particular analytical framework chosen in developing the IWG SC-CO$_2$ estimates is based on probability-weighted present value. Although this framework is appropriate for its application in regulatory impact analysis, it is not the only framework relevant to decision making under uncertainty in the context of national and international climate policy. The Intergovernmental Panel on Climate Change (Heal and Millner, 2014; Kunreuther et al., 2014) has described at length the advantages and disadvantages of alternative decision-making frameworks under uncertainty (including those cited above) for setting a carbon tax, determining a cap on carbon emissions, or employing other policy approaches.

The IWG's purpose in calculating the SC-CO$_2$, however, is to provide estimates of the net damages from emitting 1 metric ton of CO$_2$. The SC-CO$_2$ estimates will be combined with other benefit and cost calculations for a regulation that affects CO$_2$ emissions—such as an energy efficiency standard for electric appliances. The uncertainties in the estimates of other regulatory benefits and costs with which the SC-CO$_2$ will be combined have been computed using the expected net benefit approach, which forms the basis for regulatory impact analysis under EO 12866 and OMB Circular A-4. For this reason, the committee also followed that approach.

Assigning Probabilities to Inputs and Outputs

Following the expected net benefit approach requires assigning probabilities to the outputs of each module.[5] In general, the information exchanged between modules will be in the form of a distribution for each year (or other period) to facilitate Monte Carlo analysis: a frequency distribution, probability density function, or a set of values and associated probability weights that is representative of an underlying distribution. Chapter 3 outlines possible approaches to projecting future GDP, population, and emissions, using both the extrapolation of historical data and expert elicitation (see below). This approach would yield a set of GDP, population, and emissions projections that can be viewed as a representative sample from an underlying distribution. These values can then be used in the climate module, which generates, for each projection, a distribution of values of global mean temperature and other climate variables. For each socioeconomic projection and draw from the distribution of climate variables, the damages module can compute a distribution of damage estimates. Thus, the overall framework for SC-CO$_2$ estimation will have to be designed to support the large number of simulations that may be required for uncertainty analysis.

Uncertainty in the rate of per capita GDP growth can be reflected in the manner by which damages are converted to a present value in the dis-

---

[5]For details on implementation, see Chapters 3-6.

counting module (see Chapter 6). In addition to this relationship between discounting uncertainty and uncertainty in observable variables (i.e., per capita GDP), discounting also often entails ethical judgments that are not reducible to a probability distribution. This additional variability in discounting is instead assessed through sensitivity analysis (see Chapter 6).

It would also be possible to use sensitivity analysis with respect to the probability distributions that are passed from one module to another, particularly those for which uncertainties are difficult to fully specify probabilistically. Chapter 3 describes an approach that could be used in the near term to derive a joint probability distribution over GDP, population, and emissions. One could use alternate probability distributions over these variables to explore the sensitivity of SC-CO_2 estimates to the probability distribution used.

EXPERT JUDGMENT IN SC-CO_2 ESTIMATION

Construction of any model requires some form of expert judgment to make choices among alternative functional forms, input variables, or other aspects of model structure that are consistent with available data and theoretical understanding. The effects of alternative choices on model results can be particularly important when extrapolating from the conditions under which a model is estimated or calibrated (which are necessarily conditions that have been observed) to the conditions relevant to estimating the SC-CO_2, which may be far in the future and involve climates, technologies, and other factors much different from those that have been observed.

"Expert elicitation" (or "structured expert elicitation") is a method that can often prove useful in developing distributions over uncertain parameters or variables whose values need to be projected into the future. It is a formal process in which experts report their individual subjective probability distributions for an uncertain quantity. The committee believes that, for input variables having a limited empirical or theoretical basis for quantification of projections and their uncertainty, expert elicitation conducted according to best practices provides a useful and necessary approach. Expert elicitation is a method to characterize what is known about a quantity; it does not add new information as an experiment or measurement would. Ideally, it captures the best judgments of the people who have the most information and deepest understanding of the quantity of interest. For some quantities, there may be so little understanding of the factors that affect their magnitude that informed judgment is impossible or can produce only unreasonably wide bounds. Appendix C describes in detail methods for conducting expert elicitation.

PROCESS OF UPDATING THE ESTIMATES

Current U.S. government practice is vague regarding when and how a process of reviewing and updating the SC-CO$_2$ estimates might occur, which makes it difficult for stakeholders and researchers to anticipate future reviews and potential SC-CO$_2$ updates and to plan for the process. A regularized, institutionalized process would allow both groups to align their activities more sensibly. If the SC-CO$_2$ estimates are to reflect advances in scientific understanding of the climate impacts of greenhouse gas emissions and the economic impacts of climate change, a process is needed to assure that the SC-CO$_2$ estimates are updated on a regular basis. Regularizing the frequency of updates would help focus the attention of researchers on providing useful inputs to the SC-CO$_2$ process and would make the timing of updates predictable to agency staff and stakeholders.[6]

Because the SC-CO$_2$ is used in regulatory impact analyses for regulations that are being issued on a regular basis, the frequency of updates should balance the desire to incorporate improved scientific information with the need to allow for proper review of any changes. The frequency of updates needs to be short enough so that estimates of the SC-CO$_2$ do not lag too far behind the science while being long enough to allow significant new information to be generated and incorporated by the IWG and to allow for scientific peer review of the revised methods and estimates themselves. Moreover, the rate of scientific progress is variable and changes over time and is different for the many disciplines and fields involved. Overall, there is a need to balance the value of a regularized and predictable process with one that is rigidly prescribed.

RECOMMENDATION 2-4 The Interagency Working Group should establish a regularized three-step process for updating the SC-CO$_2$ estimates. An update cycle of roughly 5 years would balance the benefit of responding to evolving research with the need for a thorough and predictable process. In the first step, the interagency process and associated technical efforts should draw on internal and external technical expertise and incorporate scientific peer review. In the second step, draft revisions to the SC-CO$_2$ methods and estimates should be subject to public notice and comment, allowing input and review from a broader set of stakeholders, the scientific community, and the public.

---

[6]Although the committee does not recommend that updates to the SC-CO$_2$ be tied to the release of assessments from the Intergovernmental Panel on Climate Change or the National Climate Assessment of the U.S. Global Change Research Program, it would be desirable for the IWG to take account of new evidence included in these assessments, as well as to communicate its information needs to those groups.

In the third step, the government's approach to estimating the SC-CO$_2$ should be regularly reviewed by an independent scientific assessment panel to identify improvements for potential future updates and research needs.

This process is illustrated in Figure 2-2. Step 1 involves the technical interagency process of updating SC-CO$_2$ estimates, taking into account recommendations for improvement from the scientific community and the public, scientific advances, as well as both internal government and external technical support and scientific peer review of individual modules to ensure that the proposed improvements accurately reflect evolving evidence and approaches. Incorporation of relevant external technical support and peer review of particular components (e.g., by experts in each of the module areas), and the overall framework and implementation, prior to public notice and comment would help ensure the scientific reliability and credibility of the estimates. The result of Step 1 would be a

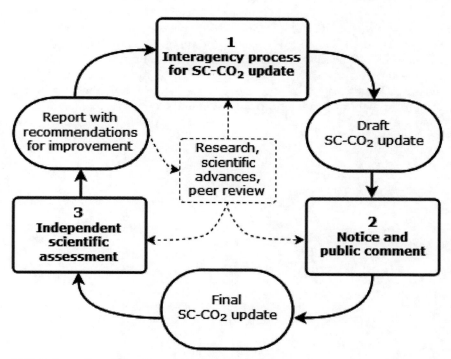

FIGURE 2-2 Regularized process for SC-CO$_2$ updates.
NOTE: See text for discussion.

draft update of SC-CO$_2$ methods and estimates. The committee estimated this step could take 2 to 3 years.

Step 2 involves obtaining input and comment on the draft update from both the public and the broadest possible scientific and technical communities, as well as other stakeholders. The result of Step 2 would be a finalized SC-CO$_2$ update for regulatory use that has incorporated public comments on the draft approach and estimates. The committee estimated this step could take 6 months to 1 year.

Step 3 involves a thorough independent scientific assessment of the SC-CO$_2$ estimation process, in order to track and assess new scientific literature over time and make recommendations for future improvements and research. The committee estimated this step could take 18 months to 2 years. The dotted box and lines at the center of the process indicate the multiple opportunities to incorporate research and scientific advances in the SC-CO$_2$ estimation process and for independent reviews to help inform research priorities.

The committee anticipates an overall process of roughly 5 years, which would allow 2-3 years between recommendations for improvements from an independent scientific assessment (end of Step 3) and the issuance of a draft SC-CO$_2$ update for public notice and comment (end of Step 1). Following from this recommendation, the committee has structured some of our other recommendations to distinguish those that we believe can be accomplished in the near term (2-3 years) from those that we believe will likely take longer to accomplish (i.e., "longer-term"). It is important that implementation of the research recommendations (in other chapters) proceed in parallel with the updating process described above.

3

Socioeconomic Module

In this chapter, the committee discusses preparation of the socioeconomic and emissions inputs to SC-CO$_2$ estimation and recommend improvements to the current IWG procedure. The chapter presents a basis for evaluating current and potential future approaches and the desired characteristics of a socioeconomic module. It also includes a survey of the resources available for the task, including scenario databases, models of the economy and emissions, means of extracting information from historical data, and expert elicitation. An illustration of an improved method for projecting population, economic activity, and emissions that could be applied in the near term, with a focus on characterizing uncertainty in the variables to be used in the climate and damages modules is provided. For the longer term, recommendations are offered for the development of a socioeconomic projection model designed to meet the special requirements of SC-CO$_2$ projection, noting that it is best supported by a program of research and development (R&D) on economic modeling frameworks.

BASIS FOR EVALUATION

The purpose of a socioeconomic module is to provide a set of projections of population, and gross domestic product (GDP) that drive projections of CO$_2$ and other relevant emissions, which are inputs to the calculation of the baseline climate trajectory. These projections take into account possible future mitigation policies and other drivers of change (see Box 2-2, in Chapter 2). The baseline emissions in turn influence the response of the climate to a pulse of CO_2 emissions (see Chapter 4). Estimates of population and GDP, perhaps disaggregated by region and sector, are also direct inputs to the damage calculations (see Chapter 5). The trajectory of GDP per capita is also needed for the committee's recommended discounting procedure (see Chapter 6).

The socioeconomic component of the current IWG SC-CO_2 estimation methodology is based on five scenarios of population, GDP, and emissions through 2100: they were selected from those produced by the detailed-structure integrated assessment models (IAMs) used in the EMF-22 multimodel comparison exercise of the Energy Modeling Forum (Clarke et al., 2009). Four of these scenarios are reference scenarios (no mitigation policy) that roughly span the distribution of reference fossil fuel combustion and industrial CO_2 emissions in the EMF-22 project. They entail atmospheric CO_2 concentrations between 612 and 889 ppm in 2100. One of the five scenarios involves atmospheric stabilization at a radiative forcing equivalent to 550 ppm CO_2 by 2100, and thus assumes moderately strict mitigation measures. The IWG extended each of these scenarios to 2300 to capture the persistence of climate change and its associated net damages, assuming that growth rates of population and per capita GDP in each scenario decline linearly to zero in 2300. The IWG does not offer a rationale for these growth assumptions.

The five scenarios used by the IWG do not span uncertainties in relevant variables (e.g., GDP, population, and energy) or reflect the broader scenario literature (e.g., Kopp and Mignone, 2012; Rose et al., 2014b). In estimating the SC-CO_2, these five scenarios are weighted equally, thereby treating them as equally likely. The IWG does not provide a justification for this implicit assumption. As discussed throughout this report, good scientific practice requires that key variables and associated uncertainties be clearly identified, characterized, and supported; that the methods used to produce probabilistic projections be consistent with the available peer-reviewed literature; and that the projections themselves be consistent with the main features of the historical record.

For estimating the SC-CO_2, the socioeconomic module needs to produce projections far enough into the future to capture the vast majority of discounted damages.[1] The committee recognizes that this may entail projecting GDP, population, and emissions two to three centuries into the future, which presents a significant challenge. Although projecting the impact of a change in radiative forcing on mean global temperature involves parametric uncertainty (see Chapter 4), the basic physics of the climate system are well established. In contrast, models that project

---

[1]"Vast majority" is a deliberately vague term that signals much more than majority, but not 100 percent.

population or GDP are subject to the behavior of individuals and social systems, which are more malleable than the principles governing physical systems. Therefore, a near-term approach for a socioeconomic module that relies on projecting historical data, combined with elicitation of expert judgment is presented. The importance of conducting sensitivity analyses for the distribution of GDP, population, and emissions to investigate their impact on estimates of the SC-CO_2 is also discussed.

For any long-term projection of population and GDP, associated projections of emissions of CO_2 from fossil fuel and industrial sources and land use change, as well as other greenhouse gases and aerosols, will depend on the joint evolution of various technologies and policies aimed at mitigating emissions. Thus, it would be desirable for the socioeconomic module to explicitly take into account the likelihood of these future changes. The committee discusses a near-term approach consistent with these criteria below ("Developing a Socioeconomic Module in the Near Term").

Two additional desirable criteria are more difficult to satisfy. The first deals with disaggregation of global totals. As discussed further below, historical experience and expert judgment provide a basis for computing a probability density function for both global average per capita GDP growth over time and for global population that are consistent with alternative economic growth projections. However, the empirically based literature on climate-related damages is typically concerned with particular regions and even particular sectors (e.g., agriculture) in each region. Unfortunately for modeling purposes, the relative contributions of different sectors and regions to global growth has varied significantly over time. For instance, in 1960 it would have been difficult to predict the rise of the Chinese economy or the fall of the Soviet Union over the following half century or the advance of computer and communications technologies and their spinoffs. In a world of many regions and many sectors, rigorous characterization of uncertainty regarding their relative contributions to global growth would require construction of a probability density function over many variables, extending far into the future—a task well beyond the current capacity of the research community. Accordingly, a less ambitious approach is recommended in the near term.

The second desirable but difficult criterion is the incorporation of feedbacks from the damages and climate modules to income, population, and emissions projections. As discussed in Chapter 2, there are many potential linkages and feedbacks between modules. Identifying the most important feedbacks and incorporating them in a fully integrated socioeconomic-climate-damages framework would represent a significant advance beyond the current state of the art.

Development of such a framework might start with the climate system impacts on human and natural systems described by Working Group

II of the Fifth Assessment Report of the Intergovernmental Panel on Climate Change (IPCC, 2014a), which identifies regions and sectors where such interactions appear to cause the most significant physical impacts. For some impacts a next step could be to incorporate recent research that assigns economic values to such impacts (e.g., Diffenbaugh et al., 2012; Reilly et al., 2012a, 2012b; Taheripour et al., 2013; Baldos and Hertel, 2014; Grogan et al., 2015; Diaz, 2016). As discussed in the final section of this chapter, such an effort would also be an important component of a longer-term research strategy.

> **RECOMMENDATION 3-1** In addition to applying the committee's overall criteria for scientific basis, uncertainty characterization, and transparency (see Recommendation 2-2 in Chapter 2), the Interagency Working Group should evaluate potential socioeconomic modules according to four criteria: time horizon, future policies, disaggregation, and feedbacks.

- Time horizon: The socioeconomic projections should extend far enough in the future to provide inputs for estimation of the vast majority of discounted climate damages.
- Future policies: Projections of emissions of CO_2 and other important forcing agents should take account of the likelihood of future emissions mitigation policies and technological developments.
- Disaggregation: The projections should provide the sectoral and regional detail in population and GDP necessary for damage calculations.
- Feedbacks: To the extent possible, the socioeconomic module should incorporate feedbacks from the climate and damages modules that have a significant impact on population, GDP, or emissions.

The next section discusses the scholarly resources that are available to construct an improved socioeconomic module in an SC-CO_2 framework. The subsequent two sections cover an approach to producing improved estimates in the near-term and a recommended longer-term strategy.

LITERATURE AND METHODS

There are four resources that can be used in the construction of socioeconomic modules: detailed-structure models, scenario libraries, time-series analysis of historical data, and elicitation of expert opinion.

Detailed-Structure Models of the Economy

The models used to generate the scenarios (available in the various libraries discussed below) are significant resources available to the IWG.[2] Among these are detailed-structure models that attempt to model the structure of the global economy. These represent nations and aggregate regions and their interaction through international trade and disaggregate the sectors that make up the individual economies. They differ from reduced-form models like the IAMs used to produce estimates of the SC-CO$_2$, SC-IAMs. SC-IAMs model a single global economy or a small number of regions and include more limited economic sectoral detail than a detailed-structure model.[3]

These detailed-structure models differ from one another in mathematical form, but they tend to fall into two general categories, partial equilibrium and general equilibrium. Partial equilibrium formulations represent particular sectors in detail (e.g., energy, agriculture) but do not consider interactions among sectors and interactions with the macro economy. Therefore, many prices in the economy are assumed to be exogenous. Examples of this type of detailed-structure IAM include the global change assessment model (GCAM)[4] and Prospective Outlook on Long-term Energy Systems (POLES) (Kitous, 2006). In contrast, general equilibrium formulations consider the market transactions and linkages among sectors (including capital, labor, resource markets, and international trade), and all treat prices in the economy as endogenous. Examples of this approach include the anthropogenic emission prediction and policy analysis (EPPA) model (Chen et al., 2015), MERGE (a model for estimating the regional and global effects of greenhouse gas reductions) (Blanford et al., 2014), and World Induced Technical Change Hybrid Model (WITCH) (Bosetti et al., 2006).

These types of models have been used not only for scenario construction, but also for more formal uncertainty analysis of energy and emissions (e.g., Reilly et al., 1987; Manne and Richels, 1994). Recently, an analysis by Gillingham and colleagues (2015) used the EPPA, GCAM, MERGE, and WITCH models (along with two reduced-form IAMs, DICE [Dynamic Integrated Climate-Economy model] and FUND [Framework for Uncertainty, Negotiation and Distribution model]) in a study that considered uncertainty in population and GDP. Another analysis (Bosetti et al., 2015) imposed uncertainty in the cost parameters of key technolo-

---

[2]An example is the set of models that contributed to the Fifth Assessment Report's Working Group III scenario database (Intergovernmental Panel on Climate Change, 2014b, Annex II, Table AII.14).

[3]For an overview of modeling terminology, see Box 2-1 in Chapter 2.

[4]See http://jgcri.github.io/gcam-doc [January 2017].

gies in the GCAM and WITCH models, while holding population and GDP constant. To generate a projection of emissions for a study of uncertainty in climate, Webster and colleagues (2008, 2011) introduced both types of uncertainty in the EPPA model, considering both uncertainty in population and drivers of GDP and uncertain distributions of many input parameters, such as elasticities, resource stocks, and technology costs.

These models produce information of use in damage estimation, including both regional and sectoral detail (e.g., the role of the agricultural sector). Moreover, many are formulated to provide additional details needed for climate modeling, such as emissions of land CO_2 and non-CO_2 greenhouse gases and their geographical distribution. Because of their formulation, these models ensure consistency in the relationships among population, GDP, and emissions of CO_2 and other greenhouse gases in each nation or region and in their aggregation to global emissions. Moreover, these features are preserved in the construction of probabilistic scenarios or other representations of uncertainty. At the same time, use of a detailed-structure model does not reduce the underlying information requirement associated with projecting regional and sectoral detail far into the future.

Scenario Databases

A number of multidecade to century-scale scenarios have been developed and have been catalogued in study-specific and scientific community assessment libraries. In 1992, the Intergovernmental Panel on Climate Change (IPCC) developed a set of global greenhouse gas emissions scenarios for use in climate change policy assessments, called the integrated scenarios 1992 (Leggett et al., 1992; Pepper et al., 1992). Through a long and complex process, the IPCC updated those scenarios in its *Special Report on Emissions Scenarios* (Nakicenovic et al., 2000).

Since then, the IAM community has published a large number of scenarios, most of them generated by specific intermodel comparison studies, some with publicly available scenario libraries. In its *Fourth Assessment Report* and *Fifth Assessment Report*, Working Group III of the IPCC assembled the research community's scenarios into large libraries in support of their respective reports (Intergovernmental Panel on Climate Change, 2007b, 2014c). The IPCC scenario libraries are a rich scientific resource with large numbers of scenarios (e.g., more than 1,000 in Intergovernmental Panel on Climate Change, 2014c), but one that needs to be used with care (see discussion below). Riahi and colleagues (2016) describe the vast amount of scenario work that has been completed, providing useful information to support future scenario construction.

Since the 2000 IPCC compilation, two specific sets of scenarios have been produced—representative concentration pathways and shared socioeconomic pathways. The former were designed to provide consistent, standardized radiative forcing information for the purpose of coordinated experiments for climate modeling (Moss et al., 2010; van Vuuren et al., 2011). The latter were designed to complement the former with additional information beyond radiative forcing to support studies of climate change impacts, adaptation, and vulnerability. Specifically, the shared socioeconomic pathway scenarios provide macro socioeconomic information (O'Neill et al., 2014), such as population structure, education levels, extent of urban development, and income distributions.[5]

Although these processes have provided much needed benchmark scenarios for coordinating the work of the various global change research communities, neither the representative concentration pathway nor the shared socioeconomic pathway scenarios was designed with SC-CO$_2$ computation in mind. More specifically, neither was formulated to characterize climate change or socioeconomic uncertainty. The broader existing scenario libraries do reflect some degree of both model and parametric uncertainty because a substantial number of modeling groups participated in these efforts. However, the libraries are problematic as the basis for developing probability distributions of population, income, and emissions because they do not formally consider parametric uncertainty or uncertainty over a full range of model input assumptions. In addition, in order to be useful, oversampling would need to be addressed in some fashion, with some models and studies represented more than others, and a variety of vintages of single models sometimes included. Meaningful statistics cannot be readily derived from these libraries without attention to these issues, even though attempts are regularly made to do so (e.g., by Working Group III of the IPCC [Intergovernmental Panel on Climate Change, 2014c]). Furthermore, scenario libraries, including the shared socioeconomic pathways, do not provide sectoral disaggregation, and they also typically extend only to 2100, even though projections beyond 2100 are important determinants of current SC-CO$_2$ estimates.

Another missing element in current scenario libraries is the effect of mitigation policies. As noted above, projections of emissions conditional on population and GDP logically need to account for the effect of future changes in mitigation policies in the United States and abroad, and such changes are themselves uncertain. Historical observations and scenario

---

[5]Standardized policy assumptions, also referred to as shared policy assumptions, were also developed to represent ways that countries might reduce greenhouse gas emissions and move from a shared socioeconomic pathway baseline to a combination shared economic and representative pathway scenario (Kriegler et al., 2014).

libraries do not on their own provide a basis for attaching probabilities to future policies. Finally, preliminary work with historical data on the global economy (discussed below) indicates that the range of economic growth rates in existing scenario libraries is too narrow to properly reflect historical experience.

In short, largely because they were not designed specifically to facilitate the computation of the SC-CO$_2$ or characterize the global-level of uncertainty in that computation, existing scenarios are not well suited for this purpose. However, as we discuss below, they may be helpful in disaggregating projections of global population and GDP to regional or sectoral scale.

Using Historical Data and Expert Judgment
for Long-Term Economic Projections

Scenarios are intended to provide an internally consistent description of a potential future, conditional on initial conditions and structural assumptions about economic system dynamics. In contrast, forecasts describe the likely future of one or more quantitative variables, often implicitly or explicitly probabilistic, based on empirical modeling.

As noted above, the IWG's analysis indicates that projections to around 2300 may be necessary to adequately represent the damages expected to result from a pulse of CO$_2$ emissions. Unfortunately, the literature contains only a few examples of projections of population, GDP, and emissions of any sort beyond 2100 and provides little discussion of how to construct them (see further discussion below). In fact, the scenario libraries do not necessarily span even the range of historical experience. For example, among the IPCC baseline scenarios that extend to 2100 and were used by Working Group III in the *Fifth Assessment Report* (Intergovernmental Panel on Climate Change, 2014c), the range of GDP growth rates is 1.1-2.5 percent (with only 1 of 263 below 1.2% and only 2 out of 263 above 2.4%). Yet the historical data show that a set of representative rates would span a significantly wider range.

A study by Mueller and Watson (2016) provides a mathematically rigorous method for using historical data to construct probabilistic growth forecasts over future time horizons that are a large fraction (or even a multiple) of the length of the historical record. This method, like any based on historical data, rests on the assumption that the stochastic process of future growth will be the same as in the past. In addition to this assumption, such methods cannot detect or incorporate fluctuations that occur over periods longer than the historical record.

The key insight of Mueller and Watson is that low-frequency, persistent variation in historical data is the most relevant information for understanding long-term uncertainty. In contrast, high-frequency, idio-syncratic variation in growth rates—for example, idiosyncratic shocks that arise each year—will average out over long horizons and will thus contribute little variability in the long run. Isolating these low-frequency variations and transforming the estimates of low-frequency contributions to growth back to the original sample space allowed the researchers to produce a representation of the low-frequency, persistent variation and use it to project a distribution of long-term average growth rates. In their work, Mueller and Watson looked at the solvency of the U.S. Social Security Trust Fund, with forecast example horizons of 75 years, using U.S. datasets as short as 67 years.

For an example relevant for the SC-CO_2 estimates, the committee used the Mueller and Watson approach for time horizons of 90 and 290 years (e.g., 2010-2100 and 2010-2300) for projections of per capita GDP growth using alternate datasets of 60 and 140 years.[6] The assumption that the stochastic process governing future growth rates will be the same as in the past is very strong, especially over such a long time ratio of projection to experience, so it would be sensible for projections produced by this or any other time-series method to be evaluated by experts before being used in SC-CO_2 analysis.

Ultimately, this approach seems most useful for informing projections of economic growth, rather than population or emissions. Population projections involve complex trends in fertility and mortality and may need to be conditioned on per capita GDP. Emissions projections without accounting for any mitigation policy can generally gain less from historical data, since there has historically been little scientific and policy attention to climate change. However, using historical emissions information to develop a no-mitigation projection might be a useful input to an expert elicitation of future emission projections, which is the subject of the next section.

Given the state of scientific knowledge and historical data, it will also be necessary to rely on expert judgment in developing a socioeconomic module. As discussed in Chapter 2 and in more detail in Appendix C, there are best practices for eliciting expert judgments about the probability distributions of uncertain quantities. As discussed below, it will be impossible to avoid reliance on expert judgment in both the near term and longer term. In most cases, the committee believes it will not be sufficient for the IWG to rely only on its own expertise. It is important to be able to draw effectively on outside experts in the relevant disciplines.

---

[6]The committee's projections involved ratios of the length-of-projection horizon to historic sample that ranged from 2.0 to 4.8, compared with 1.1 in Mueller and Watson.

DEVELOPING A SOCIOECONOMIC
MODULE IN THE NEAR TERM

As discussed above, the committee does not believe it will be possible in the near term to produce a module satisfying all of the criteria in Recommendation 3-1. However, the existing literature and methods do provide a basis for overcoming several shortcomings in the current IWG procedure. This section describes and recommends an approach that the IWG could implement in the near term.

The committee's approach is based on the assumption that important aspects of future trends will be like those in the past, with elicitation of expert opinion being the only practical way to relax that assumption. Although the ideal modeling system for SC-CO$_2$ analysis would include structured feedbacks from climate and damages to economic activity and possibly even population,[7] the committee does not believe that it is possible to build such a system in the near term. Hence, our approach for a near-term strategy, as in the current IWG approach, does not include those feedbacks. This section details the four steps in the proposed approach: (1) use econometric analysis to project economic growth; (2) develop probabilistic population projections; (3) use expert elicitation to produce projections of future emissions; and (4) develop regional and sectoral projections. It is important that this process reflect judgments as to the influence of future policies on the evolution of key technologies.

This approach also reflects the committee's view that it is advantageous to have a small number of possibly interrelated projections of population, GDP, and emissions to pass to the climate module. A small number increases transparency and facilitates expert elicitation conditional on each projection. Three values are used in this approach, which is the smallest number that both introduces variability and provides a midpoint. For example, using three projections each of population and economic growth would require the experts to generate nine probabilistic projections of emissions. If terciles are used, so that three emissions projections that can be treated as equally likely are generated for each of the nine population/GDP scenarios, this will produce 27 global-level (populations, GDP, and emissions) scenarios to be passed to the climate module and, in disaggregated form, to the damage module.

An important question is whether a single set of scenarios from the socioeconomic module should be used in the climate module or whether sensitivity analyses should be conducted by using alternate sets of scenarios. Sensitivity analysis with respect to the discount rate, for which ethical

---

[7]However, the DICE model currently used by the IWG does adjust global GDP and the capital stock for aggregate monetary climate damages each time period (see Chapter 5).

and policy considerations are relevant in addition to observable rates of discount is presented in Chapter 6. Sensitivity analysis is an appropriate way to account for ethical and policy considerations, which are especially difficult to reduce to probability distributions. In contrast, economic and population growth are observable and so a probabilistic projection approach based on historical data is appropriate for them. Emissions projections fall somewhere in between, because while historic emissions are observable, future emissions are subject to considerable policy influence. Overall, given the difficulty in projecting future GDP, population, and emissions, it would be valuable to examine the impact of alternate sets of scenarios to investigate their impact on estimates of the SC-CO$_2$.

The approach recommended in this report nonetheless focuses on a probabilistic approach to all uncertainties other than discounting. This pragmatic approach is based on the committee's recognition that there are a quite limited number of sensitivities that can reasonably be expected to be carried through a regulatory impact analysis, in which the SC-CO$_2$ is only one of many variables. In the current approach of the IWG, for example, scenarios were used for socioeconomic variables, but they were ultimately collapsed to an average by assuming equal weights on each scenario. The committee believes the recommended approach provides a better scientific basis for the assignment of probabilities to alternative scenarios.

Use Econometric Analysis to Project Economic Growth

As discussed above, recent work by Mueller and Watson (2016) examined how to estimate probability distributions of long-term growth rates in economic variables from historical data. That is, by looking at a small number of low frequency cosine transformations of historic growth rates, a predictive density of average growth rates can be constructed over an arbitrary horizon. Expert elicitation can then be applied to determine how likely the historical pattern is to hold over alternative horizons. The key underlying assumption is that behavior over the observed historical sample is a valid basis for projections over the chosen horizon. Although their application focused on the United States, using 60 years of data to construct projections over 75 years, it is straightforward to apply the same approach to global data over alternative (much longer) horizons.

As an example of such an application, the committee used data from the Maddison Project[8] to construct two time series of economic growth.

---

[8]The Maddison Project, begun in 2010, promotes and supports cooperation between scholars to measure economic performance for different regions, time periods, and subtopics. For details, see: http://www.ggdc.net/maddison/maddison-project/home.htm [October 2016].

One is a measure of growth in global GDP per capita from 1950 to 2010. Prior to 1950, data are available for only a subset of the global economy, so 1950-2010 represents the only sample for which global growth is measured. For a measure of per capita GDP growth from 1870 to 2010, we used the subset of 25 countries in Barro and Ursua (2008a, 2008b): these countries collectively accounted for 63 percent of global GDP in 1950, but for only 46 percent of global GDP by 2009.

The estimation results are summarized in Table 3-1. For additional details on the data construction and the committee's use of the Mueller-Watson approach, see Appendix D. For the 1950-2010 sample, a mean annual growth rate of 2.2 percent for real GDP per capita and a 90 percent probability interval of 0.3-4.0 percent for growth for 2010-2300 is estimated. For the 1870-2010 sample, the mean annual growth is 1.4 percent and the 90 percent probability interval is −0.8 ± 3.2 percent. The prediction intervals grow slightly, but not by much, for the longer 300-year horizon relative to 100 years.

It is unclear whether the longer series is a better basis for long-term growth projections, or the shorter series with more coverage. The longer series contains more information about long-term variation, but there are more measurement issues in the distant past so it may be less relevant for understanding behavior in the future. Even if global economic data did exist for several past centuries, for example, would one look to those data to model future uncertainty? The shorter dataset is more geographically complete, as well as more consistently measured. However, selecting the key economic jurisdictions in 1870 necessarily excludes countries that underwent transitions—through above average economic growth—into

TABLE 3-1 Estimated Annual Growth Rates Using the Mueller and Watson Procedure (in percent)

| | 2010-2100 | | 2010-2300 | |
Results	Mean Prediction	90 Percent Prediction Interval	Mean Prediction	90 Percent Prediction Interval
Results using global GDP per capita, 1950-2010	2.1	(0.6, 3.6)	2.2	(0.3, 4.0)
Results using GDP per capita measured across a subset of 25 countries, 1870-2010	1.4	(−0.4, 2.8)	1.4	(−0.8, 3.2)

NOTE: See text and Appendix D for discussion and details.

key economies in 1950 and 2010. Both kinds of estimates could be informative in selecting or creating economic growth scenarios in the SC-CO$_2$ process, as well as for inputs to expert elicitation.

After developing probability distributions for average economic growth rates over one or more horizons through statistical analysis of historic data or other means, it is desirable to translate them into a small number of projections of economic activity. The committee believes this is important for both transparency and tractability. It is easier to communicate a smaller number of discrete growth rate possibilities. It is also useful for connecting economic projections with population and emissions projections that involve expert elicitation conditional on the economic projections.

The approach discussed above would be to select representative growth rates for several equally likely fractile ranges. The example below is based on the distribution underlying Table 3-1 and using the mean of each tercile. However, one could also explore matching the standard deviation or other features of the data.

Continuing with the example calculation, Figure 3-1 shows the full cumulative distribution function for the projected average growth rate over 300 years in the example discussed above, using 1950-2010 data.

Based on this result, one can identify three terciles to use as equally likely projections, formed by breaking the cumulative distribution into three parts on the vertical axis: below one-third, between one-third and two-thirds, and above two-thirds. This division corresponds to growth rates below 1.95 percent, between 1.95 and 2.45 percent, and above 2.45 percent, as defined by the two vertical dashed lines. One can then com-

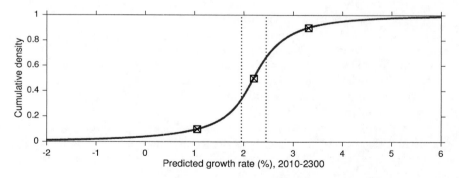

FIGURE 3-1 Sample cumulative distribution of global per capita economic growth rates.
NOTE: See text and Appendix D for discussion and details.

pute the mean growth rate in each tercile: 1.0 percent in the first, 2.2 percent in the second, and 3.3 percent in the third, indicated by boxed x's on the cumulative distribution function. These three scenarios, defined in terms of growth rates, can then be translated into projections of per capita economic activity by applying them to an initial year value.

Though this careful examination of the historical experience provides a sound basis for projection over coming decades, it may seem implausible to assume it would hold for centuries into the future, in part because of population aging or resource constraints. Thus, it will be useful for the IWG to elicit the opinions of economists and other experts concerned with long-term trends and structural change about how the length of time that such projections can be treated as representative and equally likely and how they could best be adjusted to take account of these longer-term influences. Estimates of the extent of difference with the past experience could be elicited, and the statistically derived distribution modified accordingly.

Develop Probabilistic Population Projections

Projections of population growth can take advantage of its underlying dependence on fertility and mortality rates and the age structure of society. These rates follow patterns, and the study of demography has sought to examine how these rates and the population age structure evolve over time. The International Institute for Applied Systems Analysis (IIASA) currently provides probabilistic population projections through 2100 (Lutz et al., 2014), as does the United Nations (2015a, 2015b). Both sets of projections are based on a review of the drivers of fertility and mortality in different parts of the world and (differing) judgments of what can be expected in the future (e.g., Gerland et al., 2014). For example, IIASA's central growth rate projection from 2015 to 2100 is 0.18 percent, with an 80 percent prediction interval of –0.18 to +0.51 percent.[9] Neither of these two sources report complete probability densities. It would be desirable for IIASA and the United Nations to make available the underlying probabilities, from which a small number of (perhaps three) projections could be chosen to approximate the probability density functions when treated as equally likely.

For population projections to 2300, the United Nations (2004) has published high, medium, and low projections, and Basten and colleagues (2013) have published projections under a range of assumptions about

---

[9]For the total population sheet, see http://www.iiasa.ac.at/web/home/research/ l researchPrograms/WorldPopulation/Reaging/2007_update_prob_world_pop_proj.html [October 2016].

fertility. Based on the more recent methodology (United Nations, 2015b), the probabilistic projections to 2100 could be extended further into the future. The IWG could explore that task with IIASA, the United Nations, and other researchers. Such extrapolation, like the economic projections beyond 2100, raise significant questions about whether the assumptions used in the model will hold over more than a century. It will be useful for the IWG to elicit the opinions of a group of expert demographers to validate and adjust probabilistic population projections beyond 2100.

There are reasons to expect that per capita income growth and population growth may be related in the long term. For example, more rapid rates of global economic growth would seem likely to hasten the demographic transition to lower birth rates in developing nations. Yet it seems unrealistic to expect a default inclusion of such relationships in any projections at this time given the dearth of academic research on integrated probabilistic projections of population and economic activity. Such projections could be included if the expert elicitation in economics or demography indicate the value of those relationships.

Combining population projections with each of the growth rates of per capita income would yield a relatively small set of projections of population and GDP that can be treated as equally likely and representative of the corresponding joint probability density function.

Use Expert Elicitation to Produce Projections of Future Emissions

The SC-CO_2 estimates are intended to be used in U.S. regulatory impact analysis (RIA) of proposed regulations and other policy initiatives. Accepted practice for benefit-cost analysis and the Office of Management and Budget (OMB) guidance for conducting RIAs establish that benefits and costs ought to be defined in comparison with a clearly stated alternative or "baseline," with the baseline chosen to represent what the world would be if the proposed action (i.e., program, regulation, law) is not adopted. For example, OMB Circular A-4 (p. 15) states:

> This baseline should be the best assessment of the way the world would look absent the proposed action. The choice of an appropriate baseline may require consideration of a wide range of potential factors, including:
>
> - evolution of the market,
> - changes in external factors affecting expected benefits and costs,
> - changes in regulations promulgated by the agency or other government entities, and
> - the degree of compliance by regulated entities with other regulations.

The committee notes that the consequences of any individual U.S. policy action affecting CO_2 emissions will take place in the context of other actions in the United States, as well as actions by other countries. Under uncertainty, an appropriate distribution of baselines will therefore include a range of possible outcomes for these uncertain policy developments, combined with uncertain economic and technology conditions. Thus, the committee believes the IWG acted correctly in considering scenarios with alternative levels of future global CO_2 emissions mitigation, but that SC-CO_2 estimation can be improved by making such consideration more systematic.

Although knowledge of historical experience can inform judgments about the joint evolution of various technologies and of national policies to mitigate emissions, the committee believes it would be unwise to rely heavily on statistical analysis of the sort discussed above. Instead, the committee believes there is no real alternative to relying on the judgment of experts with knowledge of both political and diplomatic processes in the United States and other nations and of technical challenges to reducing emissions.

In applying expert elicitation, as discussed in Appendix C, it would be useful for expert judgments to be informed by historical data and information about the emissions trajectories associated with different levels of climate stabilization. For each scenario of population and GDP and each greenhouse gas considered, the experts could be shown several emissions projections to provide context for their own judgment. For example, they could be shown a trajectory of emissions to 2100 consistent with extrapolation of historical experience. Such a trajectory might be obtained by projecting the historical rate of decline of CO_2 emissions per dollar of real GDP, perhaps modified by the national pledges under the Paris Agreement. They could also be shown as an emissions trajectory consistent with stabilization of CO_2 concentrations at an aggressive target level.

Having seen a range of possibilities, the experts could then be asked to provide their mean emissions projections for 2100 for that scenario, along with quantities designed to enable construction of a probability distribution. A probability density function could be created by combining the experts' judgments. And then three representative and equally likely emissions levels for 2100 could be created, and emissions trajectories could be derived from them by assuming, for instance, a constant rate of growth. Alternatively, particularly for long-lived gases such as CO_2, it may be better to work with total emissions over the period to 2100 rather than the rate of emissions at that date.

It is less straightforward to determine what useful and credible information about the period beyond 2100 could be provided to the experts. Projections of historical trends would likely be useful, although because of increasing uncertainty about technologies and policies, they are likely to be less useful than for the period to 2100. Emissions projections under the assumption of strict abatement would also likely be useful. In eliciting judgments for both the periods before and after 2100, allowance would also need to be made for the possibility that net emissions will go to zero, with a range of uncertainty around the dates involved.

These first three steps of the four-step procedure suggested above will yield a relatively small set (e.g., on the order of 27 members) of global population/income/emissions scenarios that are representative of the underlying probability density functions. These results can then be used in the climate module (discussed in Chapter 4) to produce inputs to the damage module.

In the committee's approach, it is essential that the socioeconomic module pass emissions projections of other climate significant forcers to the climate model. However, because asking experts to produce representative trajectories of other climate forcers for each of nine or more population/income scenarios would be unduly burdensome, simplified procedures are likely to be required. It may be sufficient to ask experts to deal with only a few of the most important forcing agents or only a few extreme scenarios and to use interpolation or other simple methods to produce the desired inputs. Whatever simplified procedures are adopted, however, it would be best if they are based on expert judgments and be clearly described and the rationale for adopting them explicitly presented.

Develop Regional and Sectoral Projections

Damage calculations are likely to require projections of population by region as well as projections of GDP by region and sector. These details will likely be needed in the calibration of aggregate climate damage functions and as inputs to regional and sectoral damage formulations. This is no small task, particularly as one would expect such disaggregated projections to depend on specific global values and be subject to considerable uncertainty.

In this section, the committee considers three approaches to using currently available models and results to develop regional and sectoral projections for the near term. Specifically, the possibility of using scenario libraries, an individual model, or the existing SC-IAMs to develop shares is discussed. Each has advantages and disadvantages, and none of the approaches enables characterization of the uncertainty in the disaggregation step itself. In the case of population, the possibility of using existing regional and national population projections is also discussed.

The first approach would be to estimate median GDP shares for each identified region, over time, using a particular scenario library. As discussed above, the committee does not recommend continuing the IWG procedure of using such scenarios as the basis for global-level projections, but many scenario libraries do contain internally consistent projections of regional shares, generally to 2100. One can examine a collection of socio-economic scenario results and derive the population and GDP shares over time of each consistently defined region. This analysis would produce a range of share estimates from which medians could be computed for each region and time period, although such median shares might need to be rescaled to sum to 1.0 (retaining their relative weights). These adjusted median shares could then be applied to the global population and GDP projections to construct regional population and GDP projections.

For example, suppose one of the global scenarios involves a global GDP projection of $200 trillion in 2050. If the rescaled median U.S. and Chinese shares were 20 percent and 25 percent, respectively, in 2050, one would use $40 and $50 trillion as the 2050 projections of U.S. and Chinese GDP. Depending on the breadth of the scenario library, the analysis could be broken into groups of scenarios based on different underlying global population and economic levels, with the above approach applied separately to these groups. This approach would allow the disaggregation to vary across the global projections (as well as over time) in the socioeconomic module. For projections beyond 2100, extension of share projections would be required and need to rely on additional assumptions. A simple choice would be that regional shares remain constant at their 2100 values; alternatively, trends prior to 2100 (e.g., 2080-2100) might be projected to continue in some way.

An advantage of this approach to disaggregation is that it is not tied to any particular model. The median share across models is a robust measure that remains relatively unchanged as individual models are added to or deleted from the analysis. It also provides a potential mechanism to vary disaggregation across global growth projections. However, choices about near-term damage modeling may require regional resolution beyond what is available in scenario libraries, so there is the potential need for additional disaggregation. Larger libraries in particular (e.g., IPCC) tend to have low regional resolution (e.g., five global regions), as well as the sampling issues discussed above.[10] This issue highlights the need to decide which library to use. In addition, using the median share for each region does not ensure consistent shares or shares representative of a single scenario, either across regions or across population and GDP projections, in contrast to shares produced by a single model. Finally, inconsistencies might also arise if the regional shares are coupled with sectoral shares coming from another source. Scenario datasets do not currently provide sectoral disaggregation. Therefore, a different source

---

[10]The shared socioeconomic pathways dataset is another resource with 5- and 32-region resolution for some variables, but it is based on a limited number of models and projections.

is required to provide the sectoral detail that may be needed for damage calculations.

The second approach to disaggregation would be to use the baseline projection of an existing detailed-structure economic model. A time profile of regional GDP and population shares could be derived and applied to the global aggregates. Sectoral GDP shares in each region could be constructed on the basis of value added by sector. The same type of extrapolation discussed above would likely be necessary to extend the projections beyond 2100, as such models are typically limited to a 100-year horizon. A key advantage of this approach is that a model with explicitly defined regional and sectoral economic activity ensures consistency (conditional on the model's structure) among regional and sectoral activity. One disadvantage is that the regional and sectoral detail—while more extensive than the scenario libraries—still may not match the regions and sectors in the damage formulations. Another drawback is that the approach relies entirely on one model, although this disadvantage could be lessened by choosing a model that produces regional shares similar to those in the first approach.

The third approach is to derive shares from the sectoral GDP and/or regional population and GDP assumptions in the existing SC-IAMs or from other models used in the near-term updated damages module. This approach has the advantage of using information already at the appropriate level of disaggregation and properly defined for each damage formulation. The principal disadvantage is a dependency not only on one model, but also on one model of a specific subset of models. In addition, there is no mechanism to vary the path of shares over different global growth paths.

Disaggregated population projections could also be drawn directly from the source of the population projections (e.g., the United Nations, IIASA) as part of the global population projection process. In particular, the United Nations (2015b) provides country and regional probabilistic projections that could be used to develop regional projections consistent with the set of global projections. However, it would then make sense to extract GDP per capita by region from the source of regional economic detail—a scenario library or single model—rather than GDP shares. A time series of regional GDP per capita estimates could then be combined with regional population estimates to produce a new series of regional GDP projections. These projections could then be used to construct shares to disaggregate the global GDP projections. Importantly, this approach would preserve the relative GDP per capita across regions coming from the source economic modeling. This approach would presumably provide more credible disaggregated population projections, but it would require a much more involved process to couple those projections with disaggregated economic estimates.

It is important to note that most of the approaches discussed here do not simultaneously provide a consistent disaggregation of global GDP and population, match exactly the assumptions and level of disaggregation in the damages module, and rigorously consider how disaggregation is likely to vary over alternative global projections. Moreover, as noted at the outset, most of the approaches do not consider how to model the uncertainty associated with disaggregated results. The longer-term approach discussed below is designed to address these and related issues.

Given the several possible approaches and their various strengths and weaknesses, the IWG will need to compare the options to justify its proposed near-term approach. This involves a choice of how to balance the consistency of the disaggregation, the robustness of multiple models, the alignment with damage module aggregation, and the ability to capture variation across alternate global projections. Furthermore, given the possibility of using multiple damage formulations with different regional and sectoral levels of aggregation, the IWG may need to develop custom approaches for generating disaggregated input projections for different damage formulations.

RECOMMENDATION 3-2 **In the near term, to develop a socio-economic module and projections over the relevant time horizon, the Interagency Working Group should:**

- **Use an appropriate statistical technique to estimate a probability density of average annual growth rates of global per-capita GDP. Choose a small number of values of the average annual growth rate to represent the estimated density. Elicit expert opinion on the desirability of possible modifications to the implied projections of per capita GDP, particularly after 2100.**
- **Work with demographers who have produced probabilistic projections through 2100 to create a small number of population projections beyond 2100 to represent a probability density function. Development of such projections should include both the extension of existing statistical models and the elicitation of expert opinion for validation and adjustment, particularly after 2100. Should either the economic or demographic experts suggest that correlation between economic and population projections is important, this could be included.**

- Use expert elicitation, guided by information on historical trends and emissions consistent with different climate outcomes, to produce a small number of emissions trajectories for each forcing agent of interest conditional on population and income scenarios.
- Develop projections of sectoral and regional GDP and regional population using scenario libraries, published regional or national population projections, detailed-structure economic models, SC-IAMs, or other sources.

A LONGER-TERM STRATEGY AND AGENDA
FOR RESEARCH AND DEVELOPMENT

Meeting the desired features of the socioeconomic module laid out at the beginning of this chapter is a substantial challenge, though the modifications in procedure recommended in the preceding section would bring the SC-CO_2 framework closer to them. Even with these improvements, however, dependent as they are on scenario libraries and economic models developed for purposes other than SC-CO_2 estimation, shortcomings will remain.

For example, under the approaches suggested above it will be difficult to maintain consistency between regional and sectoral disaggregation of GDP and estimates of emissions, even given assumptions regarding mitigation policies. Also, it is a challenge to ensure consistency between estimates of CO_2 emissions and emissions of other greenhouse gases, such as methane and nitrous oxide. Potential feedbacks are another shortcoming. Monetary damages imply a reduction in economic activity and productive investment, reducing concurrent and future economic performance and affecting emissions net of policy as well. It is an effect illustrated in Figure 2-1 (in Chapter 2), but not considered in the current IWG procedure or in the method described in the preceding section.

In addition, understanding the net damage of climate change may require an elaboration of the four-module structure of Figure 2-1, to take more explicit account of phenomena such as climate effects on biological productivity and land use, changes in regional water availability, and the implications of human adaptation to rising temperatures and all of its associated impacts. And, most challenging, the methods and models used to prepare socioeconomic projections tend to be focused on the current century, whereas projections into subsequent centuries are required for the SC-CO_2 estimation.

In the longer term, there are many advantages to investing in the construction of a dedicated socioeconomic projection framework. Considering its unique objectives, a detailed-structure economic model designed for the task will likely be the most effective approach in the short run. An existing detailed-structure model might be applied more or less "as is" to this task, as suggested above for near term regional or sectoral disaggregation.

However, such an approach has severe limitations for the longer term. Existing detailed-structure models were formulated to meet very different objectives than those of the IWG. Many of these models support greater sectoral and regional detail than likely is needed or desirable for the SC-CO$_2$ calculation, and yet they may not yield projections of the particular variables that are needed for climate damage analysis. Feedbacks of some climate impacts have been incorporated in studies using some of these models (e.g., Reilly et al., 2012a), but these were one-time studies of particular effects. The existing models have not been configured for efficient accounting of the wider set of feedbacks that may emerge from a damage module. And, as has been noted, none of the existing detailed-structure models was designed to produce projections beyond 2100, nor does any of them provide a consistent link to other projection methods for the post-2100 period. Hence, although the existing models could play a useful role in the near term, for the longer-term what is needed is a model specifically built for that purpose.

> **RECOMMENDATION 3-3** In the longer term, the Interagency Working Group should engage in the development of a new socioeconomic module, based on a detailed-structure model, that meets the criteria of scientific basis, uncertainty characterization, and transparency, is consistent with the best available judgment regarding the probability distributions of uncertain parameters and that has the following characteristics:
>
> - provides internally consistent probabilistic projections, consistent with elicited expert opinion, as far beyond 2100 as required to capture the vast majority of discounted damages, taking into account the increased uncertainty regarding technology, policies, and social and economic structures in the distant future;
> - provides probabilistic regional and sectoral projections consistent with requirements of the damage module, taking into account historical experience, expert judgment, and increasing uncertainty over time regarding the regional and sectoral structure of the global economy;
> - captures important feedbacks from the climate and damage modules that affect capital stocks, productivity, and other determinants of socioeconomic and emissions projections.

It should enable interactions among the modules to ensure consistency among economic growth, emissions, and their consequences; and

- is developed in conjunction with the climate and damage modules, to provide a coherent and manageable means of propagating uncertainty through the components of the SC-CO$_2$ estimation procedure.

Development of such a framework, designed to satisfy the long-term needs of SC-CO$_2$ estimation, would represent an advance in economic modeling. Though an effort to build a detailed-structure model suitable for SC-CO$_2$ estimation could usefully build on one or more existing models, it would be best if supported by a program of research on economic modeling frameworks and model development.

CONCLUSION 3-1 Research on key elements of long-term economic and energy models and their inputs, focused on the particular needs of socioeconomic projections in SC-CO$_2$ estimation, would contribute to the design and implementation of a new socioeconomic module. Interrelated areas of research that could yield particular benefits include the following, in rough order of priority:

- Development of a socioeconomic module to support damage estimates that depend on interactions within the human-climate system (e.g., among energy, water, and agriculture, and between urban emissions and air pollution).
- Use of econometric and other methods to construct long-run projections of population and GDP and their uncertainties.
- Quantification of the magnitude of feedbacks of climate outputs and various measures of damages (e.g., on consumption, productivity, and capital stocks) on socioeconomic projections, based in part on existing detailed-structure models.
- Development of detailed-structure economic models suited to projections that are consistent over very long time horizons, in which functional form and levels of regional and sectoral detail in inputs and outputs may differ between the nearer term (e.g., to 2100) and the more distant future.
- Development of probability distributions of uncertain parameters used in detailed-structure models, with a particular focus on the differences among developed, transitional, and low-income economies. Examples of uncertain parameters include key elasticities of substitution (e.g., between labor

and capital inputs to production, between energy and nonenergy demand, and among fuels in total energy use), energy technology costs and rates of technology penetration, and rates of capital turnover.

There are costs as well as benefits of the committee's recommended approach to improved socioeconomic projections. Developing an SC-CO$_2$ estimation framework with a more tightly integrated socioeconomic module will take time—likely more than the 2-3 years that this report defines as the near term. Thus, some version of, or alternative to, the near-term strategy presented here will need to be used for the next revision of the SC-CO$_2$, and perhaps for one or more of the subsequent revisions.

In addition to initial model development, continual maintenance will be required to update underlying datasets and incorporate modifications to the SC-CO$_2$ procedure. Though such a dedicated model could be documented in the peer-reviewed literature, many judgments regarding its use and updating would fall to the IWG itself. It is the view of the committee that such an investment in tools to support SC-CO$_2$ estimation is warranted.

4

Climate Module

In a modular SC-CO$_2$ framework, the primary purpose of the climate module is to take the outputs of the socioeconomic module (such as emissions of CO$_2$ and other climate forcing agents) and estimate their effect on physical climate variables (such as a time series of temperature change) at the spatial and temporal resolution required by the damages module. Thus, it must (1) translate CO$_2$ (and other greenhouse gas[1]) emissions into atmospheric concentrations, accounting for the uptake of CO$_2$ by the land biosphere and the ocean; (2) translate concentrations of CO$_2$ (and other climate forcing agents) into radiative forcing; (3) translate forcing into global mean surface temperature response, accounting for heat uptake by the ocean; and (4) generate other climatic variables that may be needed by the damages module. Those other variables may include regional temperature, regional precipitation, statistics of weather extremes, global and regional sea level, and ocean pH. In so doing, it must accurately represent within a probabilistic framework the current understanding of the climate and carbon cycle systems and associated uncertainties. Figure 4-1 provides a detailed conceptual view of this module.

The committee's proposed climate module can draw on a rich scientific literature regarding the physical behavior of the Earth system. Models for projecting climate change have evolved from a few equations

---

[1]CO$_2$ is not the only important climate forcing agent; other key agents include methane, nitrogen oxides, fluorinated gases, and aerosols. To accurately estimate the response of the climate system to a pulse release of CO$_2$, any Earth system model needs to include the effects of these other agents as well, as the response depends nonlinearly on climate itself.

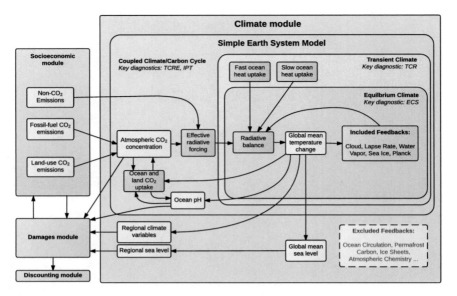

FIGURE 4-1 Conceptual view of the internal flow of the climate module.
NOTES: Output variables are shown in yellow. The list of excluded feedbacks is shown. See text for discussion. See Box 4-1 for definitions of TCR, TCRE, IPT, and ECS.

using planetary energy balance to estimate global mean surface temperature changes, to Earth system models of intermediate complexity and full complexity Earth system models that project coupled changes in the atmosphere, oceans, and land surface.[2] At each stage of the development of Earth system models, more comprehensive representations of feedbacks and response characteristics have been added (Flato et al., 2013), leading to increases in model resolution and the extent to which the complexity of the Earth system is represented in model structures. These representations have built on knowledge about mechanisms and relationships gleaned from increasingly comprehensive and longer-term observations of the Earth system (see, e.g., National Research Council, 2012).

Modern Earth system models represent the physics, chemistry, and biology of the atmosphere, oceans, and terrestrial hydrosphere and bio-

---

[2]The intermediate complexity models share the structure of full complexity models, but they have a reduced set or a parameterized set of processes and feedbacks that allows faster model runs and exploration of uncertainty.

sphere at spatial and temporal scales that allow representation of their interactions and feedbacks. While energy balance models have global- or hemispheric-mean spatial resolution and annual-mean time steps, state-of-the-art Earth system models have ~100 km or finer resolution in the atmosphere and land and ~25 km resolution in the ocean with 15-minute time steps. With additional components and increasing model resolution, Earth system models capture most key elements of the scientific community's current understanding of the complex coupled dynamical systems that govern both the Earth system's internal variability in the absence of forcing and its response to external forcing agents.

Any SC-CO_2 estimation framework has to account for uncertainty in projections of both global mean surface temperature changes and related climate variables. Computational demands of full complexity models and even of intermediate complexity models limit their ability to provide this kind of probabilistic information when very large ensembles of model runs over very long time horizons are required, as is the case with the estimates for the SC-CO_2. Hence, SC-CO_2 calculations require computationally efficient simple Earth system models that represent the critical behaviors captured in more comprehensive models and account for the key sources of uncertainty in climate projections. Implicitly, this also requires that such simple Earth system models are capable of reproducing key observational climate records of the past few centuries.

The next section discusses the general characteristics of a useful Earth system model, and the third section provides an overview of a simple Earth system model that satisfies these criteria. The following four sections cover key elements of that system: sea level rise; ocean acidification; spatial and temporal disaggregation, through estimating higher resolution climate variables from simple low-resolution models; and uncertainty propagation. The committee then considers some limitations of common approximations made in simple Earth system models, and the chapter concludes with a discussion of research needs.

CHARACTERISTICS OF AN ADEQUATE CLIMATE MODULE

The committee's Phase 1 report (National Academies of Sciences, Engineering, and Medicine, 2016) suggested several criteria that could be used to evaluate whether any simple Earth model considered for use in SC-CO_2 estimation reflects current scientific understanding of the relationships between CO_2, other greenhouse gases, emissions, concentrations, forcing, and global mean surface temperature change, as well as their uncertainty and profiles over time. These criteria are reiterated and expanded in Recommendation 4-1.

RECOMMENDATION 4-1 In the near term, the Interagency Working Group should adopt or develop a climate module that captures the relationships between CO_2 emissions, atmospheric CO_2 concentrations, and global mean surface temperature change, as well as their uncertainty, and projects their profiles over time. The module should apply the overall criteria for scientific basis, uncertainty characterization, and transparency (see Recommendation 2-2 in Chapter 2). In the context of the climate module, this means:

- Scientific basis and uncertainty characterization: The module's behavior should be consistent with the current, peer-reviewed scientific understanding of the relationships over time between CO_2 emissions, atmospheric CO_2 concentrations, and CO_2-induced global mean surface temperature change, including their uncertainty. The module should be assessed on the basis of its response to long-term forcing trajectories (specifically, trajectories designed to assess equilibrium climate sensitivity, transient climate response and transient climate response to emissions, as well as historical and high- and low-emissions scenarios) and its response to a pulse of CO_2 emissions. The assessment of the module should be formally documented.
- Transparency and simplicity: The module should strive for transparency and simplicity so that the central tendency and range of uncertainty in its behavior are readily understood, reproducible, and amenable to improvement over time through the incorporation of evolving scientific evidence.

The climate module should also meet the following additional criterion:

- Incorporation of non-CO_2 forcing: The module should be formulated such that effects of non-CO_2 forcing agents can be incorporated, which will allow both for more accurate reflection of baseline trajectories and for the same model to be used to assess the social cost of non-CO_2 forcing agents in a manner consistent with estimates of the SC-CO_2.

Comprehensive Earth system models are the scientific community's best representations of the current understanding of the many interacting components of the Earth system. However, simple Earth system models can represent the relationship between emissions, atmospheric composition, and global mean surface temperature in a manner consistent with more comprehensive models: as shown in Box 4-1, their parameters can be set to reproduce the behavior of more complex models under a range of relevant forcing scenarios. Such consistency can be evaluated using a number of coordinated benchmark experiments that have been performed with Earth system models: in the next section, several that are particularly useful in assessing simple Earth system models that are intended to be used in SC-CO_2 estimation are highlighted. Performing well against these diagnostics is not a guarantee that a climate module is appropriate for all applications, so conclusions can also, where possible, be checked against direct calculations carried out with more comprehensive models.[3]

As defined in Box 4-1, four key metrics can describe the configuration of a simple Earth system model: equilibrium climate sensitivity (ECS), transient climate response (TCR), transient climate response to emissions (TCRE), and the initial pulse-adjustment timescale (IPT). In addition, the overall response of the simple models to forcing can be assessed using the representative concentration pathway or extended concentration pathway (RCP/ECP)[4] experiments driven by total forcing (Collins et al., 2013). The key point of comparison is whether a simple model's central projections and projection ranges agree with those of more comprehensive Earth system models. These diagnostics would not necessarily disqualify models based on broader responses than Earth system models, which are known to cluster near central estimates (e.g., Huybers, 2010; Roe and Armour, 2011). Also, simple models can include feedbacks not represented in more comprehensive models because of more complex models' high computational requirements, but the diagnostics could be analyzed using runs with these additional feedbacks disabled so as to facilitate comparison with more complex models that do not include such feedbacks.

---

[3]A simple Earth system model is calibrated against more comprehensive models rather than directly against observations because there is no direct estimate of parameters such as equilibrium climate sensitivity or transient climate response because the relationship between global mean quantities in a simple model and corresponding (incompletely) observed quantities is often ambiguous (see, e.g., Richardson et al., 2016). Thus, it is generally preferred to calibrate a simple model against more comprehensive models (which have in turn been tested against observations) using idealized experiments in which, for example, only CO_2 concentrations are varied.

[4]Extended concentration pathways are an extension of RCP emissions scenarios from 2100 through 2300 (van Vuuren et al., 2011). See Chapter 3 for an introduction to the RCPs used in the Fifth Assessment Report of the Intergovernmental Panel on Climate Change.

BOX 4-1
Timescales and Key Metrics for Relating CO_2 Emissions to Temperature Change

The response of global mean temperature to climate forcing can be characterized by a number of different metrics, which represent different timescales and include different processes and feedbacks: see Figure 4-1-1.

Equilibrium climate sensitivity (ECS) measures the long-term response of global mean temperature to a fixed forcing, conventionally taken as an instantaneous doubling of CO_2 concentrations from their preindustrial levels. The "long-term" time frame is set by the time it takes for the ocean as a whole to equilibrate with the change in forcing, typically on the order of many centuries to a couple of millennia. It is a measure of long-term planetary response, but it is not comprehensive. It includes the effects of atmospheric and ocean processes involving clouds, water vapor, snow, and sea ice. However, it does not include other mostly slower processes that, at least until recently, have not been represented in coupled global climate models, such as those involving vegetation, land ice, or changes in the carbon cycle.

Transient climate response (TCR) measures the transient response of global mean temperature to a gradually increasing forcing. It is measured on a time frame that allows the shallow "mixed layer" of the ocean to approach equilibrium with the changed forcing before equilibration of the deep ocean is achieved. In models, TCR is assessed by increasing CO_2 concentrations at 1 percent per year until CO_2 concentrations double in year 70: TCR is the average temperature increase achieved by the two decades around the time of doubling (years 61-80).

Transient climate response to emissions (TCRE) measures (on a similar timescale as TCR) the ratio of warming to cumulative CO_2 emissions. Although TCRE has become a widely used metric over the past decade, it has a shorter history in the scholarly literature than the measure of ECS or TCR and so the methods for assessing it are less established. In models, one way of assessing TCRE is from experiments similar to the 1 percent per year increase used to assess TCR, but using emissions rather than a prescribed change in concentrations to drive the experiment (see, e.g., Gillett et al., 2013). TCRE is then estimated as the ratio of TCR to the cumulative CO_2 emissions at the time of CO_2 doubling.

BENCHMARK EXPERIMENTS FOR CALIBRATING AND EVALUATING SIMPLE EARTH SYSTEM MODELS

Temperature Response to Idealized Concentration Changes

The simplest benchmark experiments involve changing atmospheric CO_2 concentrations in a (simple or complex) climate model and com-

Initial pulse-adjustment timescale (IPT) has only recently been a focus of research and does not have a standard name or definition, but it may be of considerable importance for estimates of the SC-CO$_2$, which are driven by the injection of a pulse emission of CO$_2$. It measures the initial adjustment timescale of the temperature response to a pulse emission of CO$_2$ (Herrington and Zickfeld, 2014; e.g., Joos et al., 2013; Ricke and Caldeira, 2014; Zickfeld and Herrington, 2015). For example, Joos and colleagues (2013) assessed the IPT by adding a 100 gigaton (Gt) carbon pulse (367 Gt CO$_2$) to baseline emissions that stabilized CO$_2$ concentrations at a reference level of 389 ppm: the IPT from such an experiment is the time over which temperatures converge to their peak value in response to the pulse.

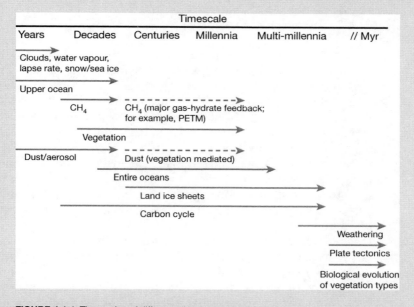

FIGURE 4-1-1 Timescales of different climate processes.
SOURCE: PALAEOSENS Project (2012, Figure 1).

puting the resulting global mean surface temperature response. Simple climate models,[5] including those used in estimating the SC-CO$_2$, have

---

[5]The committee refers to "simple climate models" here rather than "simple Earth system models" to encompass models that do not have a fully interactive carbon cycle (i.e., calculating, rather than prescribing, the distribution and fluxes of carbon within the climate model.)

traditionally used ECS as a key summary indicator of the sensitivity of the climate system to changing CO_2 concentrations. Since the 1990s, another widely used indicator has been TCR (see Box 4-1, above, for definitions). Successive reports of the Intergovernmental Panel on Climate Change (IPCC) have noted that ECS and TCR co-vary (Meehl et al., 2007) and that TCR is typically the more policy-relevant parameter (Frame et al., 2006; Otto et al., 2013). It is also better constrained by climate observations to date (Gregory and Forster, 2008; Libardoni and Forest, 2011, 2013). Because these quantities co-vary, varying ECS alone in any probabilistic assessment without checking the implied distribution for TCR risks introducing an implicit distribution for TCR that can be inconsistent with available observations (Meinshausen et al., 2009).

Figure 4-2 illustrates the concepts of ECS and TCR:

- Panel (a) shows global mean surface warming in idealized 1 percent per year increasing-CO_2 experiments performed with the Climate Model Intercomparison Project (CMIP5) comprehensive Earth system models (black lines) compared with the Fifth Assessment Report (AR5) (see Collins et al., 2013, Figure 12.45f) assessed range for TCR (red vertical bar) and the response of a simple Earth model system (blue plume).
- Panel (b) shows warming following an instantaneous quadrupling of atmospheric CO_2 concentrations in CMIP5, clearly showing a two-timescale response, with expected equilibrium warming based on assessed range for ECS (red vertical bar) and response of a simple Earth model system (blue plume).
- Panel (c) shows atmospheric concentrations in CMIP5 1 percent per year increasing-CO_2 experiments plotted against cumulative CO_2 emissions, compared with the historical observed airborne fraction (cumulative emissions and increase in atmospheric concentrations over the 1870-2011 period—diamond, with dashed line showing extrapolation), showing the consistent increase in airborne fraction with warming and cumulative emissions in complex Earth system models.
- Panel (d) shows temperatures in CMIP5 1 percent per year increasing-CO_2 experiments plotted against cumulative CO_2 emissions, showing the straight-line relationship characterized by the TCRE.

The discussion of ECS and TCR would pertain to a model driven entirely by an endogenous forcing pathway (see the boxes labeled "Equilibrium Climate" and "Transient Climate" in Figure 4-1).

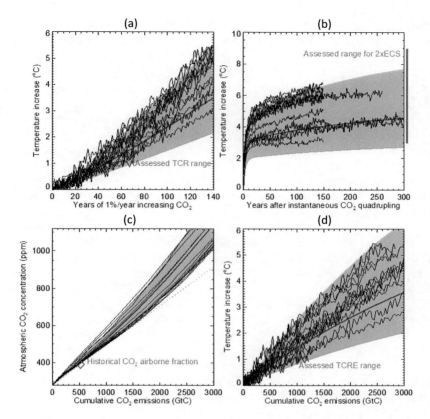

FIGURE 4-2 Examples of global mean surface warming in response to various changes in CO_2 concentrations.

NOTES: Panel (a) shows the response to an idealized 1 percent per year CO_2 increase sustained for 140 years (to quadrupling) of the CMIP5 ensemble of comprehensive climate models (black lines) and of the FAIR model (see text), with the IPCC AR5 assessed range for TCR (1-2.5 °C).

Panel (b) shows the corresponding response to an instantaneous quadrupling of CO_2 concentrations. For comparison, the IPCC's assessed range for ECS (1.5-4.5 °C) is shown, increased by a factor of two to correspond to a CO_2 quadrupling.

Panel (c) shows the relationship between diagnosed cumulative CO_2 emissions in the 1 percent per year runs and atmospheric CO_2 concentration, with the convex shape indicating an increasing airborne fraction over time.

Panel (d) panel shows diagnosed cumulative CO_2 emissions against warming, showing the approximate straight-line relationship discussed in the text.

The black lines reflect the results of comprehensive Earth system models. The blue plumes represent results from a simple Earth system model. See text for discussion.

SOURCE: Adapted from Collins et al. (2013, Figure 12.45f) and data from the. Coupled Model Intercomparison Project, CMIP5.

The solid black lines in panel (a) show the response of global mean surface air temperature in the CMIP5 Earth system models to a 1 percent per year increasing-CO_2 scenario initiated in year 1. After the initial decade or so, all models indicate an approximately straight-line increase in temperature with time, with superimposed fluctuations due to internal climate variability.

The black lines in panel (b) show the response to an instantaneous quadrupling of atmospheric CO_2 concentrations in year 1. Almost all models show a rapid initial adjustment over a decade or two, followed by a gradual warming that continues over many centuries as the global oceans slowly come into equilibrium with this new radiative forcing. Both timescales are relevant to the calculation of the SC-CO_2, with the initial adjustment timescale being primarily relevant at high discount rates and the slow longer adjustment timescale relevant at low discount rates.

The red vertical bars in panels (a) and (b) show "likely"[6] ranges of uncertainty for the transient climate response and the equilibrium climate sensitivity as assessed by IPCC AR5. In panel (b), the limits of the 1.5-4.5 °C assessed range for equilibrium climate sensitivity are doubled to 3-9 °C to allow direct comparison with the response of Earth system models to a quadrupling of CO_2 concentrations.[7] As expected, the CMIP5 model range in year 70 of these integrations (see Figure 4.2a) coincides closely with the assessed likely range for TCR. In contrast, the complex models are still far from spanning the assessed range of uncertainty in ECS even after 300 years of integration. By definition, ECS represents the warming of the climate system after it has been allowed an infinitely long time to re-equilibrate with a constant atmospheric composition, and this equilibration takes centuries to millennia in the current generation of Earth system models.

Since atmospheric composition is not expected to be constant over these timescales under any emission scenario, ECS is less directly relevant to the climate system response on policy-relevant timescales. Its prominence is to some extent a historical artefact, in that it was the aspect of the climate response that could be assessed with the "slab-ocean" climate models of the late 1970s and 1980s (e.g., Manabe and Wetherald, 1975, 1980; National Academy of Sciences, 1979; Manabe and Stouffer, 1980).

---

[6]"Likely," in IPCC terminology (Mastrandrea et al., 2010) and as used here, means that the true value has a 66 percent or higher probability of being within the quoted range. "Very likely" means the true value has a 90 percent or higher probability of being within the quoted range.

[7]Although ECS is defined as the response to a CO_2 doubling, it can be evaluated against any increase, allowing for the logarithmic relationship between the change in CO_2 concentration and the temperature response: in the CMIP5 model intercomparison, ECS was evaluated using a CO_2 quadrupling.

TCR is most relevant to the calculation of the SC-CO_2 for high values of the discount rate that emphasize the decadal response, while ECS is more important at very low discount rates in which integrated damages are dominated by the multi-century response.

The blue shaded plumes show the response of a simple Earth model system (discussed below) with low, best-estimate, and high values for TCR (panel a) and ECS (panel b). The model is consistent with the more complex Earth system models in that it reproduces key features of the model's responses, including the linear warming after the initial decade in panel (a) and the short and long timescales of response in panel (b). Therefore, any simple climate model would have to support at least two response timescales (Held et al., 2010; Caldeira and Myhrvold, 2013; Geoffroy et al., 2013).

The ranges of uncertainty (shaded plumes) are matched to the IPCC's assessed ranges for ECS and TCR shown by the red bars: they have not been explicitly fitted to the Earth system models, and indeed appear biased slightly low relative to the distribution of the models' results. This is because the IPCC-assessed ranges of uncertainty in these climate system properties are based on a number of lines of evidence in addition to these climate model results—including evaluation of recent climate change and radiative forcin110g, the recent global energy budget, and paleoclimate observations—so an exact correspondence would not be expected. Emergent properties of the climate system like TCR or ECS cannot be observed directly, so all efforts to constrain them rely on some combination of observations and (simple or complex) climate modeling, and the IPCC combines multiple approaches to provide a single assessment that is consequently more robust than any estimate based on a single study.

Relationship between Emissions and Concentrations

Since the mid-2000s, many Earth system models have incorporated interactive carbon cycles, and these idealized experiments have been extended to diagnose the emissions required to increase CO_2 concentrations at a prescribed rate, in addition to the uptake of CO_2 by land and ocean and the residual "airborne fraction." Panel (c) in Figure 4-2 shows atmospheric concentration of CO_2 in the idealized experiments shown in panel (a) plotted against cumulative diagnosed CO_2 emissions, which are the total amount of CO_2 that would need to be emitted into these models to achieve the prescribed increase in CO_2 concentrations, accounting for uptake by the land and oceans in the model's carbon cycle. The slope of these lines indicates the airborne fraction: an increase of 1 ppm in concentrations for every 2.12 gigatons of carbon (GtC) (7.77 gigatons of

CO_2 [Gt CO_2])[8] of emissions would indicate an airborne fraction of unity, meaning all CO_2 emitted remains in the atmosphere. Airborne fraction in the CMIP5 models (black lines) is initially about 45 percent, similar to that observed over the historical period (dashed line and diamond), but increases as the climates warm and CO_2 accumulates, due to the weakening of land and ocean carbon sinks (Jones et al., 2013). The lines are clearly convex (curving upwards), with the convexity accurately reproduced by the simple climate model (blue plume) discussed below.

The coupled climate carbon cycle response to emissions can be summarized in a plot of global mean surface temperature change against diagnosed cumulative CO_2 emissions from the comprehensive Earth system models included in CMIP5 under the 1 percent per year increasing-CO_2 scenario (Figure 4-2, Panel a). Despite the diversity of the CMIP5 models, the results show a linear relationship between long-term warming and cumulative CO_2 emissions for cumulative emissions up to about 2,000 GtC (7,333 Gt CO_2) (Gillett et al., 2013). More recent experiments show this approximate linearity holds in some models for cumulative emissions up to 5,000 GtC (18,333 Gt CO_2) (Tokarska et al., 2016). This approximate linearity arises from a cancellation between the rising airborne fraction and the concave (logarithmic) relationship between CO_2 concentrations and forcing. The slope of the temperature/cumulative emissions relationship is called TCRE.

Human-induced warming to date is consistent with this straight-line relationship between cumulative CO_2 emissions and CO_2-induced warming. However, the signal-to-noise ratio is low enough that it would also be consistent with other functional forms. Hence, it is difficult to predict the consequences of future emissions based simply on extrapolating a purely empirical approach. The two effects that give rise to this straight-line relationship in more complex models are both well supported by observations and theory. Reproducing the relationship, therefore, represents a minimum requirement for a simple Earth system model. It is not sufficient in itself, particularly in a model that is used to represent the response to both CO_2 and other forcings: hence the need to check the temperature response of the model to idealized concentration changes (Panels (a) and (b)) and the airborne fraction (Panel (c)). More specific experiments can also be used to ensure that a simple model is reproducing the behavior of more complex Earth system models for the correct reasons. For example, in Gregory et al. (2009) and Arora et al. (2013), warming is artificially suppressed while CO_2 concentrations increase at 1 percent per year and emissions are diagnosed as before. This allows the biogeochemistry-induced increase in the airborne fraction to be separated from the climate-induced

---

[8]Each 1 ton of CO_2 contains 0.273 tons of carbon.

increase. Verifying that a simple climate model can reproduce the relationship between cumulative emissions and concentrations under such an idealized scenario is an additional test that the changing airborne fraction (Panel (c)) is occurring for realistic reasons (Millar et al., 2016).

The spread of TCRE among Earth system models emanates from the varying sensitivities of land and ocean carbon processes to climate change, and their subsequent impact on atmospheric CO_2 concentrations and climate (Arora et al., 2013). As a result of competition between CO_2-sensitive photosynthesis uptake and temperature-sensitive respiratory release, there is little agreement about the sign or magnitude of carbon-climate feedbacks on land uptake at the end of the 21st century. The spread among Earth system models in shifts in precipitation location, amounts, and timing further compounds this uncertainty.

For the AR5 (Ciais et al., 2013), CMIP5 Earth system models with interactive land and ocean carbon cycles coupled to the physical climate system were used to infer the emissions that would be compatible with historical and representative concentration pathway (RCP) trajectories of CO_2 concentration. The uncertainty in land uptake propagates to a wide range for the "compatible" cumulative fossil fuel emissions for 2012-2100: 140-410 GtC (513-1503 Gt CO_2) for RCP 2.6 and 1415-1910 GtC (5188-7003 Gt CO_2) for RCP 8.5. Ocean uptake is more consistent than land uptake across Earth system models; however, scientific understanding of the biological carbon pump, especially in an acidifying ocean, remains rudimentary. Research into the responses of the land and carbon cycles and their changing capacities to absorb and store carbon is much needed. Much of the experimental and field research undertaken thus far has focused on the responses of the marine biota to increasing CO_2 and temperatures. To be useful for estimating SC-CO_2, the experimental design could include decreasing CO_2 and constant temperature, as may occur with a pulse release (Joos et al., 2013).

Response to a Pulse Injection of CO_2

Since the SC-CO_2 is defined in terms of the impact of a pulse injection of CO_2 into the atmosphere, one highly relevant test of the performance of a simple Earth model system is to compare its response to a pulse injection with that of more comprehensive models. This comparison is complicated by the strong dependence of the pulse response on the reference trajectory and the lack of any coordinated intercomparisons of comprehensive models focusing specifically on the pulse response to a standardized set of CO_2 and non-CO_2 forcings. The most comprehensive intercomparison study to date is that of Joos and colleagues (2013), in which a collection of Earth system models, Earth system models of intermediate complexity,

(a)   (b)

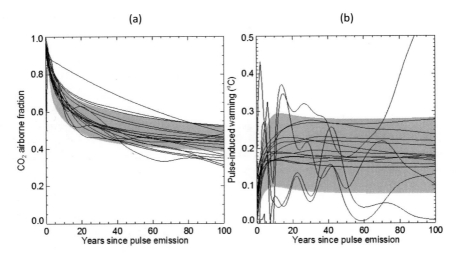

FIGURE 4-3 Fraction of injected CO_2 remaining in the atmosphere [panel (a)] and response in surface air temperature [panel (b)] to a pulse injection of CO_2 in 2015 (year 0) against a background scenario of approximately constant CO_2 concentrations from 2010.
NOTES: The figure includes a range of full-complexity Earth system models, Earth system models of intermediate complexity, and simple Earth system models (black thin lines). Dark blue thick line and blue shaded region represent the median and range and mean of the response of the simple coupled climate carbon cycle model. See text for discussion.
SOURCE: Data from Joos et al. (2013) and the Coupled Model Intercomparison Project, CMIP5.

and simple Earth system models were driven with observed CO_2 concentrations and non-CO_2 forcing to 2010, and concentrations and forcing held constant thereafter. CO_2 emissions were then diagnosed and the models were then re-run twice, once with the diagnosed emissions and a second time with a 100 GtC (367 Gt CO_2) pulse of CO_2 injected instantaneously in 2015. The difference between these latter two simulations provides a measure of the response to a CO_2 pulse.

The temperature response following a pulse injection, shown in Figure 4-3, indicates the initial pulse-adjustment timescale (IPT), which is a measure of the timescale over which temperatures converge to their peak value in response to the pulse.[9] The IPT is less than a decade in most

---

[9]Most precisely, the IPT is the timescale over which the gap between the realized temperature and the peak temperature decays to $1/e$ (~37%) of its size at the time of the pulse (i.e., the exponential decay timescale).

Earth system models, meaning that peak temperatures are reached in less than two decades. This timescale is particularly important for SC-CO_2 calculations at high discount rates because it determines how rapidly an injection of CO_2 generates impacts. The suite of models show that peak temperatures are maintained for the duration of the model integrations, ~1,000 years, at which time about a quarter of the CO_2 pulse remains in the atmosphere. As atmospheric CO_2 concentrations decrease, deep ocean temperatures adjust toward equilibrium at a similar rate (Solomon et al., 2009), stabilizing surface temperatures.

This standard "impulse-response" experiment of Joos and colleagues (2013) has the advantage that many different modeling centers have performed an identical experiment. It highlights that the models converge on the deep ocean being the larger repository of the added CO_2 on millennial timescales. In the first 100 years after a pulse release, large uncertainties are associated with the sink estimations, especially those of the land, echoing the CMIP5 model results where the sensitivities of land uptake to CO_2 and temperature have much greater spread among the models than those of ocean uptake (Arora et al., 2013; Ciais et al., 2013). These uncertainties propagate to atmospheric CO_2 concentration and the temperature response. The impulse-response experiment has the disadvantage, however, that holding CO_2 concentrations constant from 2010 means that the 100 GtC pulse is introduced into an artificial "baseline" scenario of rapidly falling emissions. More realistic impulse-response experiments with comprehensive models and research into the capacities of the land and oceans to store carbon with changing climate and emissions are discussed in the conclusions of this chapter.

Response to Historical Forcings and Future Scenarios

Another test of a simple Earth model system is to compare its behavior with that of more comprehensive models when driven with observed emissions and radiative forcing over the historical period followed by a range of future forcing scenarios, such as the RCPs (Van Vuuren et al., 2011). The four RCPs are labeled RCP 2.6, RCP 4.5, RCP 6.0, and RCP 8.5, based on their respective forcing agents (in W/m^2) from long-lived greenhouse gases in 2100: see Figure 4-4.

Reproducing the relationship among CO_2 emissions, atmospheric concentrations, and temperatures under these scenarios of differing realistic rates and magnitudes of climate forcing can be considered as a final check rather than a means of tuning parameters in a simple Earth system model, because the multiplicity of different factors contributing to realistic historical or scenario experiments means that a simple model can reproduce the behavior of a more complex model, or the real world, for

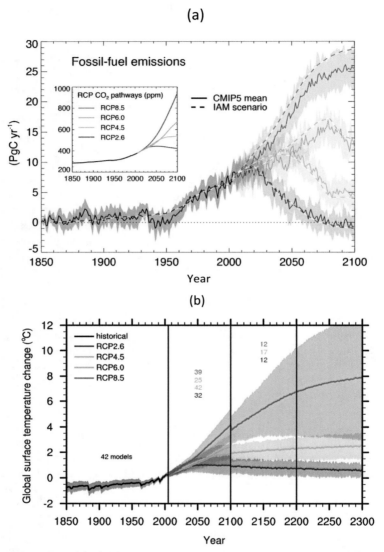

FIGURE 4-4 Fossil fuel CO_2 emissions, concentrations, and temperature response to the representative concentration pathway (RCP) scenarios as simulated by the CMIP5 Earth system models.

NOTES: Panel (a) shows the emissions, the inset shows the concentrations, which also include the response to land-use change emissions, and the right panel shows the temperature response to all human-induced climate forcing, including other greenhouse gases and aerosols. Panel (b) also shows response to extended scenarios for 2100-2300, showing long-term warming commitment.

SOURCE: Collins et al. (2013, Figure 12.5) and Ciais et al. (2013, Figure 6.25).

unrealistic reasons. The idealized experiments described above provide clearer information on a model's response to CO_2.

Numerical Distributions of Key Metrics

Periodic assessments of the literature regarding ECS, TCR, and TCRE are provided by the IPCC and can be used by the IWG. The most justifiable estimates of the probability distribution of these metrics will draw on a broad body of scientific research, and the IPCC provides a capable forum for conducting such assessments. In between IPCC assessments (which occur approximately every 7 years), it is likely that new results will be published indicating values that lie both at the high and the low end of IPCC assessed values. For example, the AR5 gave a "likely" range of 1.0-2.5 °C for TCR, based on 5-95 percent ranges from a number of different studies, while Shindell (2014) suggests that a TCR value less than 1.3 °C is "very unlikely," and Richardson and colleagues (2016) suggest an upward revision in the upper bound. Conversely, Lewis and Curry (2014), while finding a 5-95 percent range in agreement with the AR5 range, argue for a best-estimate value toward the lower end. As Richardson and colleagues (2016) demonstrate, the precise numbers can be sensitive to the choice of observations used, the assumptions underlying the analysis method, and even the definition of global average surface temperature. On a more subtle level, it has long been known (Frame et al., 2005) that statistical prior assumptions can affect the modes of an estimated statistical distribution of an indirectly observed climate parameter in ways that may not be transparent to a user. Reliance on any individual study therefore risks introducing volatility into SC-CO_2 estimates; it can be avoided by relying on the IPCC's more comprehensive periodic assessments based on multiple lines of evidence.

The AR5 provided formally assessed uncertainty ranges for ECS, TCR, and TCRE, although it does not specify either distributional forms or joint distributions. The AR5 also does not provide formally assessed ranges for other climate metrics that are relevant to the SC-CO_2 estimates, including IPT, the TCR/ECS ratio (also known as the realized warming fraction, [RWF]), and the expected increase in CO_2 airborne fraction between the 20th and 21st centuries (although this latter quantity is, to some degree, implicit in TCRE). Hence, although much of the information on the climate system response required by the IWG is contained in the IPCC assessments themselves, it would likely be necessary to consult relevant experts (including the responsible IPCC authors and reviewers) to ensure this information is used consistently. There is also an opportunity for the IWG to inform future IPCC assessments by highlighting important policy-relevant metrics on which specific guidance is requested.

The assessed IPCC likely range for ECS is 1.5-4.5 °C, while the assessed likely range for TCR is 1.0-2.5 °C. In the coupled models of the CMIP5 ensemble, ECS and TCR are strongly correlated, but TCR and the RWF are nearly uncorrelated (Millar et al., 2015). A convenient way of capturing the correlation between ECS and TCR is thus to specify TCR and RWF as a joint distribution of two statistically independent parameters; a likely range of 0.45-0.75 for RWF is consistent with the AR5 ranges for TCR and ECS. For TCRE, AR5 estimates a *likely* global warming of 0.8-2.5 °C per 1000 GtC cumulative emission for cumulative emissions less than 2000 GtC; subsequent studies (Tokarska et al., 2016) suggest the linearity may extend to a higher range, while others have found that it may not (Herrington and Zickfeld, 2014).

Although these ranges are referred to as "likely" by the IPCC, they are closer to 90 percent confidence intervals in the majority of supporting studies, and they also encompass about 90 percent of model responses in the CMIP5 ensemble. The reason for the IPCC's more conservative likelihood qualifier is that structural uncertainties, common to all studies and models, may affect conclusions. In general, there are two ways of dealing with structural uncertainty: it can be parameterized by including an additional error term, or quantitative results can be computed ignoring structural uncertainty and conclusions subsequently qualified to account for that omission. The IPCC takes the second approach, recognizing that any quantitative representation of structural uncertainties that are common to all studies and models would be difficult to justify. This is illustrated, for example, in Figure 4-5a. Consistent with the IPCC's supporting studies, 90 percent of ECS/TCR values lie in the 1.5-4.5/1.0-2.5 °C interval, so to be consistent with the IPCC's interpretation, 90 percent ranges of the outputs in the other three panels ought to be interpreted as "likely" ranges of uncertainty.

Thus, a number of methods exist to translate uncertainty ranges assessed by the IPCC into probability distributions. In the interest of transparency, the IWG could define explicitly the interpretation it proposes to use in consultation with relevant IPCC authors and reviewers. One possible option in Figure 4-5a is presented, while recognizing that others are defensible.

RECOMMENDATION 4-2 To the extent possible, the Interagency Working Group should use formal assessments that draw on multiple lines of evidence and a broad body of scientific work, such as the assessment reports of the Intergovernmental Panel on Climate Change, which provide the most reliable estimates of the ranges of key metrics of climate system behavior. If such assessments are not available, the IWG should

derive estimates from a review of the peer-reviewed literature, with care taken so as to not introduce inconsistencies with the formally assessed parameters. The assessments should provide ranges with associated likelihood statements and specify complete probability distributions. If multiple interpretations are possible, the selected approach should be clearly described and justified.

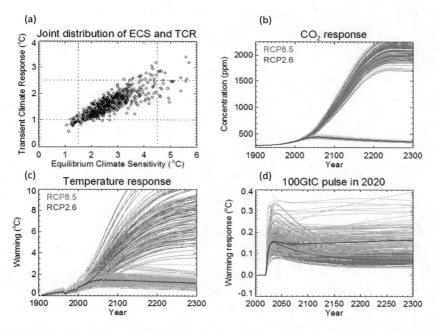

FIGURE 4-5 Demonstration of Finite Amplitude Impulse Response (FAIR) model with probabilistic sampling of key parameters.
NOTE: Panel (a) shows the joint distribution used for ECS and TCR. Panel (b) shows CO_2 concentrations in response to prescribed emissions associated with RCP 2.6 (blue) and RCP 8.5 (red/pink). Panel (c) shows temperature response to RCP 2.6 and RCP 8.5. Panel (d) shows temperature response to a pulse emission of 100 GtC released in 2020 into a background RCP 2.6 (blue) and RCP 8.5 (red/pink) scenario.
SOURCE: Adapted from Millar et al. (2016).

Transparency and Simplicity

The basic physics of the equilibrium global mean temperature response to radiative forcing has been understood since the late 19th century. More recent work has shown that the dynamics of the global mean temperature response to forcing and to emissions in complex climate models can be reproduced by simple approximations. Simple models bring great benefits in terms of both transparency and the ease with which they can be used in a probabilistic mode; thus, it makes sense for SC-IAMs[10] to use Earth system models that are as simple as possible while accurately capturing key behaviors of the climate system. Models that can be readily reproduced from a minimal set of well-documented equations are particularly useful. An example of good practice is the model provided for the calculation of greenhouse gas metrics, which is fully documented in the Supplementary Online Material of Chapter 8 of the AR5 (Myhre et al., 2013) and provides the basis for the Finite Amplitude Impulse Response (FAIR) model, detailed below.

Incorporation of non-CO_2 Forcing Agents

CO_2 is not the only important climate forcing agent; other key agents include methane, nitrogen oxides, fluorinated gases, and aerosols. To accurately estimate the response of the climate system to a pulse release of CO_2, any Earth system model needs to include the effects of these other agents as well, as the response depends nonlinearly on climate itself. This approach also allows the same modeling framework to be used for the calculation of the social cost of climate forcing agents other than CO_2. Non-CO_2 greenhouse gases generally exhibit simpler biogeochemical cycles than CO_2, and their atmospheric concentrations can be reasonably well approximated by a simple exponential decay (Myhre et al., 2013).

Aerosols are short lived in the atmosphere. While their global average climate forcing can be crudely approximated as proportional to total emissions, different spatial patterns of emissions give rise to significantly different spatial patterns of temperature change. These spatial patterns cannot be directly modeled in a simple Earth system model (see discussion of disaggregation below), so an approximation of effective forcing as proportional to emissions is reasonable, but it introduces ambiguity in the interpretation of global average aerosol forcing in the context of simple models. This ambiguity is one of the key reasons that attempting to calibrate a simple Earth system model's properties against historical

---

[10]These are the three integrated assessment models widely used to produce estimates of the SC-CO_2 (see Chapter 1).

observations using simple energy-balance models is problematic (e.g., Shindell, 2014).

AN ILLUSTRATIVE SIMPLE EARTH SYSTEM MODEL: OVERVIEW

As an example of a simple Earth system model that satisfies the criteria set forth above, the committee considered the FAIR model (Millar et al., 2016). FAIR is a minor modification of the model used in the AR5 to assess the global warming potential of different gases (Myhre et al., 2013), which the committee will call the Static Impulse Response (SIR) model. FAIR is extended with a state-dependent carbon uptake to incorporate feedbacks between the climate and the carbon cycle and thus reproduce the CO_2 behavior of more complex models, in particular the changing airborne fraction with rising temperature and cumulative emissions, which is shown in Figure 4-2c (above).

In the Earth system, the rate of CO_2 loss from the atmosphere is governed by exchange with the ocean, the terrestrial biosphere, and, ultimately, geological reservoirs. To represent this as simply as possible, FAIR divides the excess atmospheric CO_2 concentration above the preindustrial baseline value, C_0, into four fractions, denoted R_i, all of which are empty in preindustrial equilibrium. Each emission of CO_2 is partitioned between the fractions in proportions specified by a_i, and each fraction has its own loss time constant, τ_i. A single state-dependent scaling factor, α, modulates the four time constants and is defined in equation (4). CO_2 concentrations in the four fractions are updated thus:

$$\frac{dR_i}{dt} = a_i E - \frac{R_i}{\alpha \tau_i}; i = 1...4, \tag{1}$$

where E is the CO_2 emissions rate, expressed for convenience in terms of atmospheric parts per million per year (1 ppm = 2.12 Gt C = 7.77 Gt CO_2). This is mathematically equivalent to modeling the carbon cycle with four reservoirs of different capacities between which carbon is allowed to flow at different rates, although the R_i in equation (1) refer to fractions of excess CO_2 in the atmosphere (i.e., above preindustrial levels) that are responding on different timescales, and do not correspond to actual quantities in different biogeochemical reservoirs.

Atmospheric CO_2 concentrations are given by adding concentrations in the different fractions to preindustrial concentrations, $C = C_0 + \Sigma_i R_i$, and radiative forcing, F, by:

$$F = \frac{F_{2x}}{\log(2)} \log\left(\frac{C}{C_0}\right) + F_{ext}, \tag{2}$$

where F_{2x} is the forcing due to a CO_2 doubling, and F_{ext} is the non-CO_2 forcing.

For the energy balance component, FAIR estimates the temperature, T_i, for two ocean layers (i.e., thermal reservoirs) that have slow and fast response timescales (d_1 and d_2). Thus:

$$\frac{dT_i}{dt} = \frac{q_i F - T_i}{d_i}; T = \sum_i T_i; i = 1, 2. \tag{3}$$

The parameters q_1 and q_2 can be set to give any desired combination of ECS and TCR: ECS $= F_{2x}\sum_i q_i$; TCR $= F_{2x}\sum_i \beta_i q_i$, where β_i represents the fraction of the equilibrium warming of the ith response component that is manifest after a 70-year linear forcing increase, $\beta_i = 1 - d_i (1 - \exp(-70/d_i))/70$ (as described in Millar et al., 2015). The shorter of the two thermal adjustment times, d_2, largely determines the IPT (see below for representative values).

The sole structural difference between FAIR and the static impulse response model is the introduction of the state-dependent coefficient α. A suitable state-dependence for α can be determined from its relationship with the 100-year integrated impulse response function, iIRF$_{100}$, discussed in Joos et al. (2013), which is the integral of the concentration response over the century to a unit pulse emission of CO_2:

$$\text{iIRF}_{100} = \frac{1}{C_{im}} \int_{t=t_0}^{t_0+100} [C'(t) - C(t)] dt = \alpha \sum_i a_i \tau_i \left[1 - \exp\left(\frac{-100}{\alpha \tau_i}\right) \right]. \tag{4}$$

In this equation, $C'(t)$ represents the CO_2 concentration at time t following the emission pulse, C_{im}, added at time t_0, and $C(t)$ the CO_2 concentration without the pulse. FAIR assumes that iIRF$_{100}$ is a simple linear function of accumulated perturbation carbon stock in the land and ocean, which is the difference between cumulative emissions to date ("reference" emissions plus pulse) and the excess carbon in the atmosphere (i.e. neglecting geological uptake on these timescales), $C_p(t) = \int_{t'=0}^{t} E(t')dt' - (C(t) - C_0)$, and of global temperature departure from preindustrial conditions, T:

$$\text{iIRF}_{100} = r_0 + r_C C_p + r_T T. \tag{5}$$

FAIR is integrated by computing iIRF$_{100}$ at each time-step using C_p and T from the previous time-step using equation (5), computing $\square$ using equation (4) and applying it to the carbon cycle equations (1). Hence, the iIRF$_{100}$ is only exactly reproduced under constant background conditions with infinitesimal perturbations. Values of $r_0 = 35$ years, $r_C = 0.02$ years/GtC and $r_T = 4.5$ years per degree Celsius (°C), with other parameters as given in the supplementary online material of Myhre and colleagues (2013), together with ECS = 2.7 °C and TCR = 1.6 °C, give a numerically computed $iIRF_{100}$ of 53 years for a 100 GtC pulse released against a background CO_2 concentration of 389 ppm following a historical build-up. This value is consistent with the central estimate of Joos and colleagues (2013).

As noted above, the IPCC does not provide explicit distributions of ECS and TCR. Most supporting studies indicate positively skewed distributions, although not in most cases as heavily skewed as that of Roe and Baker (2007). Pueyo (2012) argues that for scaling parameters like ECS and TCR—positive quantities in which the larger the parameter, the greater the uncertainty—a log-normal distribution might be appropriate. Noting that the "likely" ranges quoted by the IPCC correspond to 5-95 percent ranges in the supporting studies, assuming a log-normal distribution for TCR with a 5-95 percent range of 1.0-2.5 °C, together with a normal distribution for RWF with a 5-95 percent range of 0.45-0.75, gives a joint distribution of ECS and TCR that is consistent with the distribution of more complex Earth system models.

Reproducing a distribution for TCRE requires accounting for the additional uncertainties in the carbon cycle. The AR5 does not provide assessed uncertainty ranges in carbon cycle properties, but varying $iIRF_{100}$ by ±7 years (5-95% range) gives a distribution of CO_2 concentration trajectories consistent with uncertainties of past emissions and concentrations (shown in Figure 4-2c, above). It also provides a 5-95 percent range for TCRE derived from 1 percent per year increasing-CO_2 experiments of 0.8-2.6 °C/TtC (Figure 4-2d, above), in close agreement with the AR5 assessed "likely" range. This plot of CO_2-induced warming against cumulative CO_2 emissions is very similar to the corresponding plot derived from more complex models (see Intergovernmental Panel on Climate Change, 2013, Figure 10), in that it is almost straight and slightly concave at high values. A simple climate model that omits carbon cycle uncertainty represented in the state-dependent $iIRF_{100}$ would necessarily display a very different shape, strongly concave over the full range.

Finally, the key parameter determining the IPT is the short thermal adjustment time, d_2. The IPCC does not give an assessed range for IPT, so a median value of 4.1 years with a 5-95 percent range of 2-7 years, based on the range of behavior of the CMIP5 models (Geoffroy et al., 2013), is used for illustration here. Results are generally insensitive to the specification of the longer timescale, d_1. For consistency, and in the absence of an assessed range for this parameter, the committee also uses the multimodel mean estimate from Geoffroy and colleagues (2013) of 229 years. Blue lines in Figure 4-5b shows the response of the global mean temperature to a pulse injection of 100 GtC of CO_2 in 2020 against a background ambitious mitigation (RCP 2.6) scenario. The overall behavior is very consistent: a rapid adjustment on a timescale of the order of a decade or less to a temperature plateau that persists for a century or more. The red/pink lines show the corresponding result against a background no-mitigation (RCP 8.5) scenario: the rapidly rising background warming results in a declining response to the input pulse.

As a proof of concept, FAIR provides an example of a model consistent with all three of the criteria in Recommendation 4-1 (above). First, its parameters can be set so as to yield distributions of ECS, TCR, TCRE, IPT, and responses to RCP pathways consistent with the responses of more complex earth system models, as illustrated in Figures 4-2 and 4-5 (above). Second, it is simple and transparent. Third, non-CO_2 radiative forcing can be straightforwardly introduced through the F_{ext} term, and because the model is structurally identical to that used by the IPCC for lifetime and metric calculations for a broad range of greenhouse gases, it can readily be applied to compute the response to, for example, methane and nitrous oxide emissions with a simple change of values of the parameters a_i and τ_i. Note that, for gases whose behavior can be characterized by a single exponential decay life-time, three of the a_i can be set to zero.

Each element of FAIR is necessary to allow ECS, TCR, TCRE, and IPT to be set independently and to demonstrate relevant behaviors seen in higher complexity models. Consistent with equation (3) (above), the climate system exhibits two dominant timescales of response to forcing, reflecting the response of the mixed layer and the deep ocean (Hansen et al., 1984; Gregory, 2000; Held et al., 2010; Geoffroy et al., 2013). Consistent with equation (1) (above), four timescales are necessary to describe the uptake of CO_2 by the land biosphere, surface ocean, deep ocean, and geological reservoirs (Joos et al., 2013).

The feedback between the climate and the carbon cycle represented by the scaling factor, α in equations (1) and (4), is necessary to yield a near-linear relationship between cumulative emissions and warming; if α is fixed to equal 1, as in the static impulse response model, this behavior cannot be reproduced (Millar et al., 2016). Similarly, this feedback is necessary to show the increase in airborne fraction needed to jointly reproduce, with a single set of model parameters, both 20th- and 21st-century behavior seen in the CMIP5 Earth system models (see Figure 4-2 above).

The comparison of the FAIR model to the benchmark experiments described above are shown in Figures 4-2 and 4-5 (above), using a representative distribution of parameters: note how both comprehensive Earth system models and the simple Earth system models show a rapid initial adjustment (short IPT) to a pulse emission in 2020. The FAIR model shows an approximately constant temperature response over the first few centuries, although its millennial timescale behavior appears to underestimate the persistence of the warming. This illustrates the importance of using such comparisons to identify aspects of simple model behaviors that are particularly relevant to the SC-CO_2 estimation: whether the model response beyond 300 years is relevant would depend on the discount rate and damage function, among other factors.

Comparing the minimal simple FAIR model described above to the simple Earth system models in the existing SC-IAMs (see Appendix E), the committee finds that each of the SC-IAM models omits at least one key element. Specifically, all SC-IAMs omit the short adjustment timescale of the thermal response (although the Dynamic Integrated Climate-Economy [DICE] supports two response timescales, as implemented, both are multidecadal or longer). DICE also omits the feedback from climate change to the carbon cycle, which would impact its long-term response, and the short carbon cycle adjustment timescale, which would impact its IPT. The Framework for Uncertainty, Negotiation and Distribution (FUND) and Policy Analysis of the Greenhouse Effect (PAGE) models both represent the thermal response with a single (multidecadal) timescale only. Carbon cycle feedbacks are represented in FUND and PAGE, but it would require further research to establish whether these representations are structurally equivalent to FAIR. With these exceptions, the model components of DICE, FUND, and PAGE are structurally equivalent to special cases of the FAIR model described above, and hence they could be modified to satisfy the criteria outlined in Recommendation 4-1 (above) and the requirements in Conclusion 4-1. Furthermore, differences in the implementation of the models that are affecting results could also be addressed.

> **CONCLUSION 4-1 The simplest possible model capable of (a) flexibly representing equilibrium climate sensitivity (ECS), transient climate response (TCR), and transient climate response to emissions (TCRE), and initial pulse adjustment timescale (IPT) and (b) incorporating responses to forcing agents other than CO_2 requires:**
>
> - **two timescales, one subdecadal, the other centennial, of the surface temperature and ocean heat content response to radiative forcing;**
> - **at least three distinct timescales of the atmospheric CO_2 response to emissions, corresponding to atmospheric exchanges with the land and surface ocean, the deep ocean, and geological reservoirs; and**
> - **a state-dependent carbon cycle in which the fraction of emitted CO_2 that remains in the atmosphere increases in response to higher temperatures and accumulation of carbon in the land and ocean.**

PROJECTING SEA LEVEL RISE

Global mean sea level (GMSL) rise is one of the key physical parameters relevant for estimating climate damages. GMSL rise results from both the transfer of water mass from continental ice sheets and glaciers into the ocean and the volumetric expansion of ocean water as it warms. Historically, direct anthropogenic transfer of water between the continents and the oceans, through groundwater depletion and the construction of dams, has been a tertiary contributor to GMSL change (Church et al., 2013).

In principle, heat uptake in a model like FAIR could be used to diagnose the contribution of thermal expansion to global mean sea level rise. However, as noted by the AR5, thermal expansion accounts for less than half of both historical and projected GMSL (Church et al., 2013), so accounting only for this term would provide an incomplete estimate of GMSL rise. A variety of authors have demonstrated methods for probabilistically projecting GMSL rise, based either on bottom-up accounting of contributing factors (e.g., Jevrejeva et al., 2014; Kopp et al., 2014; Slangen et al., 2014) or on top-down, semi-empirical, statistical estimates of the relationship between global mean temperature and global mean sea level (Rahmstorf, 2007; Grinsted et al., 2009; Vermeer and Rahmstorf, 2009; Kopp et al., 2016a). Because the different contributors to GMSL change exhibit different spatial patterns (Milne et al., 2009; Kopp et al., 2015), only bottom-up accounting directly allows for projection of local sea level changes. Starting from Rahmstorf (2007), Kopp and colleagues (2016a) demonstrate a semi-empirical model, calibrated against a 2-millennia record of temperature and sea level change, that agrees well with bottom-up estimates, including those of the AR5 (Church et al., 2013; Kopp et al., 2014), while Mengel and colleagues (2016) demonstrate a semi-empirical method, calibrated against model-based estimates of different contributing factors, that yields similar results. Both examples provide suitable models for estimating GMSL rise from global mean temperature projections.

In the model from Kopp and colleagues (2016a), global mean sea level *h* is described by

$$\frac{dh}{dt} = b[T(t) - T_e(t)] + \varphi(t) \tag{6}$$

$$\frac{dT_e}{dt} = [T(t) - T_e(t)] / \rho_1 \tag{7}$$

$$\frac{d\varphi}{dt} = -\varphi / \rho_2, \tag{8}$$

where T is global mean temperature, T_e is the global mean temperature with which sea level is in quasi-equilibrium, φ is a multi-millennial scale contribution to sea level rise from Earth's long-term climate cycles, b is a scale factor, and ρ_1 and ρ_2 are timescales. Figure 4-6 shows an example of projections from this model.

However, semi-empirical models are, by construction, calibrated to the historical record over the past couple of centuries or millennia and do not reflect novel behaviors not exhibited in this record. The emerging agreement between semi-empirical models and many bottom-up models could be interpreted as suggesting that bottom-up models also exhibit a historical bias.

Indeed, DeConto and Pollard (2016) suggest that all these projections may be underestimating the 21st century Antarctic contribution to sea level rise by excluding some important physical processes involving ice shelves and ice cliffs. In contrast to the AR5's projection of a likely –0.04 to +0.14 m contribution from Antarctica over the 21st century under RCP 8.5 (Church et al., 2013), DeConto and Pollard (2016) suggest that the physics of ice shelf hydro-fracturing and ice cliff collapse could allow contributions of 1.3 m or more. This is an emerging area of research that will require monitoring. Advances in semi-empirical models of sea level rise are qualitatively different from most new publications addressing metrics for energy balance models, such as ECS and TCR, that appear between IPCC assessments because they are incorporating physical processes that have not previously been taken into account.

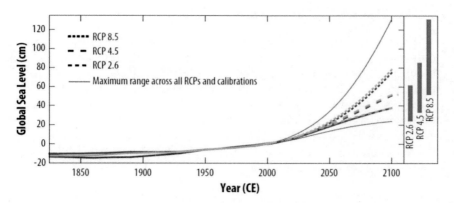

FIGURE 4-6 Projections of GMSL rise under three RCPs using the semi-empirical model of Kopp et al. (2016a).
NOTES: The black curve shows the historical reconstruction of Kopp et al. (2016a), while the nearly overlapping dashed orange and green curves show median projections under three RCPs, using temperature calibrations to either Mann et al. (2009) (orange) or Marcott et al. (2013) (green). Bars show the 90 percent probability interval of projections for 2100.
SOURCE: Kopp et al. (2016a, Figure 1e-f).

CONCLUSION 4-2 Semi-empirical sea level models provide a simple and probabilistic approach to estimate the global mean sea level response to global mean temperature change and its uncertainty. However, both semi-empirical models and many more detailed models of sea level change may exhibit a bias toward historical behaviors. In particular, they may not account for some ice sheet feedbacks and threshold responses that were unimportant over the past several millennia but could become important in response to human-induced climate change. Accordingly, estimates of sea level rise and sea level–related damages, particularly beyond 2100, need to be used with the recognition that they may understate long-run sea level uncertainty in ways that are difficult to quantify.

RECOMMENDATION 4-3 In the near term, the Interagency Working Group should adopt or develop a sea level rise component in the climate module that (1) accounts for uncertainty in the translation of global mean temperature to global mean sea level rise and (2) is consistent with sea level rise projections available in the literature for similar forcing and temperature pathways. Existing semi-empirical sea level models provide one basis for doing this. In the longer term, research will be necessary to incorporate recent scientific discoveries regarding ice sheet stability in such models.

Sea level rise is not spatially uniform, so GMSL projections may need to be regionalized for use in the damages module. Bottom-up projections of regional sea level rise (e.g., Kopp et al., 2014) can be used to calibrate the relationship between global the mean sea level and regional sea level change. A reasonable approximation of these bottom-up estimates may be represented as a linear function of global mean sea level change. A better approximation can be achieved by representing local sea level as the sum of a nonclimatic, constant-rate term and a climatic component that scales with global mean sea level change, such as:

$$SL(x,t) = \kappa(x)h(t) + m(x)t, \tag{9}$$

where SL indicates local relative sea level at location x and time t, $\kappa(x)$ a scaling coefficient, and m the rate of non-climatic processes. Figure 4-7 shows an estimate of $\kappa(x)$ and its uncertainty, as well as of $m(x)$, from Kopp and colleagues (2014). The committee notes that $\kappa(x)$ is not identically unity due to a range of factors including ocean dynamics and the

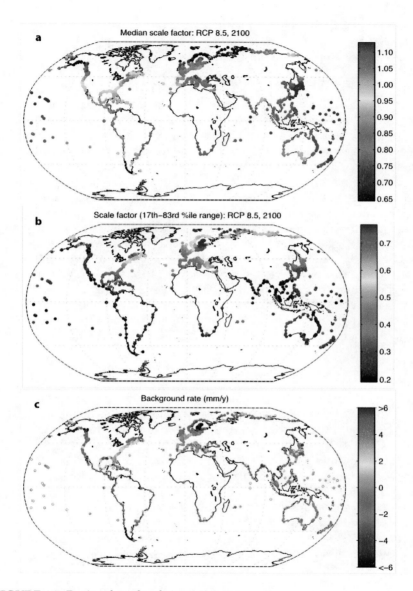

FIGURE 4-7 Regional sea level rise estimates.
NOTE: Panel (a) Median scale factor $\kappa(x)$ for the relationship between climatically driven local sea level change and global mean sea level change, panel (b) the likely (17th-83rd percentile) range of uncertainty in $\kappa(x)$, and panel (c) mean estimate of the nonclimatic rate of sea level rise $m(x)$, as estimated at a global network of tide-gauge sites.
SOURCE: Kopp et al. (2014, Figure 6).

gravitational, rotational, and land motion effects of redistributing mass between the ocean and the cryosphere (Kopp et al., 2015).

PROJECTING OCEAN ACIDIFICATION

CO_2 dissolves in seawater to form carbonic acid. As the oceans have absorbed about one-quarter to one-third of the anthropogenic CO_2 emissions, the oceans have steadily become more acidic, with pH decreasing by 0.02 units per decade since measurements began in 1980s (Bates, 2007; Doney et al., 2009; Dore et al., 2009). Ocean acidification has damaging consequences for many organisms, such as corals, bivalves, and mollusks that produce shells or skeletal structures out of carbonate minerals, as well as microorganisms at the base of the marine food web.[11] In this way, ocean acidification can contribute to the SC-CO_2 estimate through both damages to fisheries and damages to ecosystem services (Cooley and Doney, 2009; Cooley et al., 2015; Gattuso et al., 2015; Mathis et al., 2015). To the committee's knowledge, ocean acidification is included in only one integrated assessment model (IAM) (Narita et al., 2012).

Ocean carbonate chemistry is fairly well understood, and so ocean pH can be parameterized as a function of the partial pressure of CO_2 in surface waters: Figure 4-8 (see Appendix F for the derivation):

$$pH = -0.3671 \log(pCO_2) + 10.2328, \tag{10}$$

where $pH = -\log_{10}[H^+]$ is defined on the "total" hydrogen ion scale (Dickson, 1981) and pCO_2 is in micro-atmospheres. Globally averaged surface ocean pCO_2 lags behind globally averaged atmospheric CO_2 by approximately 1 year, and so the trend in pH can be readily derived from the trend of atmospheric CO_2 in simple Earth system models.

Another approach for deriving pH in simple models is as a quadratic function of concentration of dissolved inorganic carbon (DIC), with the three coefficients themselves quadratic functions of temperature (see Appendix F for the equation and its derivation). As shown in Figure 4-8 (right panel), the upper ocean becomes more acidic with increasing concentration of dissolved inorganic carbon and with increasing temperature.

---

[11]In laboratory experiments and in limited coastal studies, some commercially important shellfish species (e.g., mussels, oysters, scallops, clams, crabs) show decreased development or shell dissolution in more acidic waters (e.g., Fabry et al., 2008; Barton et al., 2012). Juveniles are particularly sensitive to acidification, and these consequences may be exacerbated by ocean warming (see, e.g., Rodolfo-Metalpa et al., 2011). The impacts of acidification propagate through marine food webs to aquaculture and marine fisheries. Furthermore, damaged coral reefs reduce tourism, coastal protection, and biodiversity.

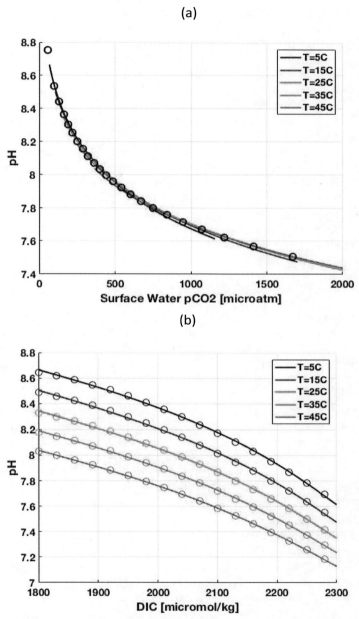

FIGURE 4-8 Variation of pH with partial pressure of CO_2 in surface waters, panel (a), and with DIC and temperature, panel (b).

NOTES: Calculated with carbon chemistry code CO2SYS.m (solid) and with empirical equations (circles). See Appendix F for the equations and their derivation.

This approach would be the starting point for estimating regional changes in pH.

The degree of ocean acidification thus is directly related to the amount of anthropogenic CO_2 taken up by the oceans as a function of time. In turn, acidification alters the relative abundance of carbonate species in surface waters and slows the ocean uptake of anthropogenic CO_2. This feedback is captured implicitly in simple earth system models whose pH projections are consistent with those in earth system models where ocean carbonate chemistry and biology is included explicitly. This is illustrated with FAIR. Atmospheric CO_2 fractions in the FAIR model do not represent actual amounts of carbon in any specific location. Rather they represent perturbations away from equilibrium for adjustments on a given timescale. If one assumes that the shortest (4-year) adjustment timescale for atmospheric CO_2 concentrations includes uptake by the near-surface oceans, then near-surface concentration of DIC can be represented by the following formula:

$$DIC = DIC_0 + \eta \left(\sum_{i=1}^{3} R_i - \frac{\sum_{i=1}^{3} a_i}{a_4} R_4 \right), \tag{11}$$

where DIC_0 is the unperturbed DIC concentration, R_4 is the perturbation concentration in the fastest-adjusting fraction, the partition coefficients α_i are as given above, and η is a proportionality constant. Following a pulse injection of CO_2 into the atmosphere, the DIC anomaly in the near surface ocean initially increases from zero over about 4 years, and subsequently varies as penetration of excess carbon to the deep ocean in proportion to the atmospheric CO_2 concentration anomaly. A proportionality constant of $\eta = 0.43$ converts the ocean carbon uptake from R (in ppm) to DIC (expressed in micromol/kg), typically used in ocean carbon observations.

Using this relationship to convert DIC into tropical pH (assuming an initial average temperature of 25 °C and initial DIC of 2030 micromol/kg, thereby giving an initial pH of 8.15), gives a simulated pH under RCP 8.5 and RCP 2.6 that compares well with Working Group 1 of AR5 Figure 6.28: see Figure 4-9b.

The global distribution of pH is not uniform: it reflects the interaction between carbonate chemistry, biology, and ocean circulation. In general, pH is lowest, ~8.10 units, in the equatorial oceans, and increases to ~8.23 units in the Arctic Ocean (Bopp et al., 2013). CMIP5 models project globally averaged pH to decrease by 0.30-0.32 units by 2100 with the RCP 8.5 scenario, and by 0.06-0.07 units for the RCP 2.6 scenario (Ciais et al., 2013). Regional changes are projected to be greatest and fastest in the Arctic and Southern Oceans, where lower salinity (from sea ice melt and increased precipitation) and enhanced carbon uptake (from greater ice-

(a)

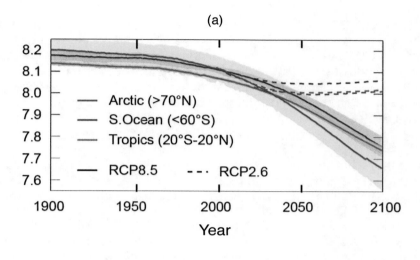

(b)

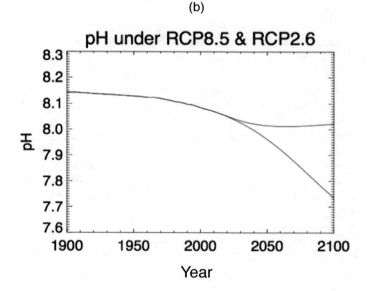

FIGURE 4-9 Projected mean change in acidification of the surface ocean from 1990 under RCP 2.5 and RCP 8.5 scenarios.
NOTES: Panel (a): estimated for the Arctic, Southern Ocean, and the Tropics by 11 earth system models (Ciais et al., 2013); panel (b): using Equation (11) in the text and FAIR (Millar et al., 2016).
SOURCE: For panel (a), Ciais et al. (2013, Figure 6.28).

free areas) exacerbate the effects of anthropogenic CO_2 uptake (Orr et al., 2005; Steinbacjer et al., 2009; Yamamoto et al., 2012). As calcium carbonate is more soluble at cold than at warm temperatures, undersaturation of aragonite, a prevalent and more soluble form of calcium carbonate, is projected to commence in the Arctic winter around 2020 and become widespread in the Arctic and Southern Oceans when atmospheric CO_2 reaches 500-600 ppm (McNeil and Matear, 2008; Steinacher et al., 2009). Coastal upwelling regions, such as the California current system, are projected to be equally vulnerable as strong seasonal upwelling brings water with higher carbon concentrations and lower pH from depth to the surface (see, e.g., Hauri et al., 2013).

Modeling of the consequences of ocean acidification on the marine biota is at an early stage, and it is mainly carried out using Earth system or regional ocean models with comparable complexity. The committee is unaware of any empirical relationship that relates regional pCO_2 or DIC changes to a projection of change in globally averaged pCO_2 or DIC: such relationships would be important for assessing regional damages associated with ocean acidification.

> **RECOMMENDATION 4-4 The Interagency Working Group should adopt or develop a surface ocean pH component within the climate module that (1) is consistent with carbon uptake in the climate module, (2) accounts for uncertainty in the translation of global mean surface temperature and carbon uptake to surface ocean pH, and (3) is consistent with observations and projections of surface ocean pH available in the current peer-reviewed literature. For example, surface ocean pH can be derived from global mean surface temperature and global cumulative carbon uptake using relationships calibrated to the results of explicit models of carbonate chemistry of the surface ocean.**

SPATIAL AND TEMPORAL DISAGGREGATION

Simple climate models produce climate projections that are highly aggregated both spatially and temporally. For example, the FAIR model produces projections of climatological (multidecadal average) global mean temperatures. Yet no one lives at 30-year global mean conditions; damages are caused by the day-to-day, place-specific experiences of the weather, the statistical properties of which are described by the climate. Thus, the damages module will either require geographically and temporally disaggregated climate variables as input or such disaggregation will need to occur in the calibration of the relationship between highly aggregated climate variables and resulting damages.

Intermediate approaches are also possible. For example, elements of the FUND and PAGE damage functions are defined with respect to climatological temperature at the spatial scale of subcontinental regions, and linear scaling relationships are used to relate global mean temperature to these regional averages. Higher-resolution climate data can be used to calibrate the relationship between regional temperature and damage. Some studies using process-based IAMs Earth system models of intermediate complexity produce latitudinal-average climate variables (e.g., Schlosser et al., 2012). Climate variables at ~1° spatial resolution and daily temporal resolution have been used to drive other studies of economic risks (e.g., Carlos et al., 2014; Houser et al., 2015; Waldhoff et al., 2015), though these have generally been bound to follow fixed scenarios (e.g., the RCPs) run with general circulation models.

The most straightforward approach to estimate the distribution of spatially disaggregated variables conditional on global mean variables is to use data from Earth system model runs to estimate linear relationships between local climate variables (e.g., temperature, precipitation) and global mean variables (e.g., temperature). This approach is known as pattern scaling (Mitchell, 2003; Tebaldi and Arblaster, 2014). With the climate module providing global mean temperature $T(t)$, the disaggregated regional climate variable D_s is estimated as a scaling by a fixed pattern, usually season dependent:

$$R_s(t,x) = T(t)p_s(x), \qquad (12)$$

where s denotes the season, x the spatial location, and p the pattern: see Figure 4-10.

Pattern scaling suffers from a number of known limitations (Tebaldi and Arblaster, 2014). It performs reasonably well for regional average temperatures; it performs less well for variables, such as precipitation, that have a high ratio of natural variability to forced change and that may have a nonlinear relationship with temperature. It also performs more reliably under conditions of rising forcing than under conditions of stable or declining forcing, as the response of the Earth system to forcing evolves over time. Some variables respond significantly differently to aerosol forcing than to greenhouse gas forcing; some slightly more sophisticated pattern scaling approaches have attempted to incorporate this dependence (Frieler et al., 2012).

Pattern scaling as generally used produces projections of climatological averages, but impact models may require higher temporal resolution. More development is needed in this area. Some researchers (e.g.,

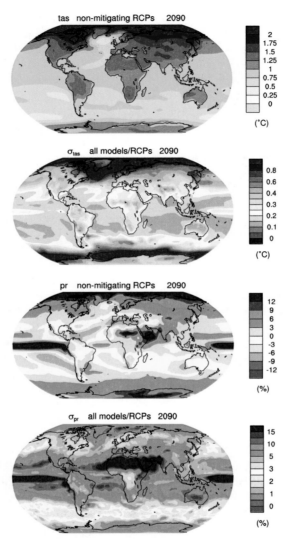

FIGURE 4-10 Mean surface air temperature and precipitation patterns for non-mitigation scenarios
NOTES: Multimodel mean patterns (panels a and c) for annual mean surface air temperature (top panels) and precipitation (bottom panels) over the 21st century for nonmitigation pathways (RCP 4.5 and RCP 8.5), and the standard deviations (panels b and d) across both models and concentration pathways (RCP 2.6, 4.5, and 8.5). Temperature maps in units of °C regional change per °C global mean temperature change; precipitation maps in units of percent regional precipitation change per °C global mean temperature change.
SOURCE: Adapted from Tebaldi and Arblaster (2014, Figures 3 and 4).

Rasmussen et al., 2016) have attempted to address this issue by combining pattern-scaled variables with time series of residuals unexplained by the linear model:

$$R_s(t, x) = T(t)p_s(x) + \tau(x, t). \tag{13}$$

These residuals could also be estimated by looking at the relationships in the observational record between climatological seasonal means and daily weather, and they could be more cleanly separated from the forced changes represented by $T(t)p_s(x)$ using large initial-condition ensembles of runs from individual Earth system models (e.g., Kay et al., 2014). Whether such temporal disaggregation is necessary in the context of the SC-CO_2 estimation depends on how the damages module is calibrated. One approach that may be most feasible in the near term, which is similar to that currently employed by the SC-IAMs, is to make the temporal disaggregation implicit in the damages module, implying that the damages module takes as input climatological average variables.

Given the available existing archives of Earth system model results, such as those produced by CMIP, one can extract data for each variable of interest for each region under different climate forcing scenarios and estimate the required scaling patterns. This approach allows one to check both the consistency across scenarios and the linearity assumptions as climate change intensifies, as well as to provide some level of uncertainty quantification based on the variance in patterns across the models. The uncertainty quantification provided by such an approach is limited, however. Ensembles of opportunity ("opportunistic samples"), such as those provided by data archives for simulations by existing climate models, are not well-formed probability distributions: the models in these archives are not independent, may underrepresent extreme outcomes, and may thus represent a biased sample of the true uncertainty in the relationship between global mean and regional variables (e.g., Tebaldi and Knutti, 2007; Sanderson et al., 2015). One solution could involve subsampling or weighted draws (e.g., Rasmussen et al., 2016). Another approach would be to produce estimates conditioned on individual models, which would be consistent with sampling across discrete distributions as suggested in Chapter 3 for baseline scenarios.

In the long run, it may be useful to use more comprehensive climate models, or statistical emulators of them (e.g., Castruccio et al., 2014), to directly estimate the joint probability distribution of global mean temperature change and regional climate changes. This approach may require a significant new emphasis in Earth system model development. Currently, most work in this area is focused on increasing the resolution and number of processes incorporated in the models. Far less work has gone into probabilistic approaches or into efforts to characterize high-impact, low-probability states of the world, but such efforts will likely be more informative for efforts to assess the SC-CO_2 and its uncertainty.

> **RECOMMENDATION 4-5 To the extent needed by the damages module, the Interagency Working Group should use disaggregation methods that reflect relationships between global mean quantities and disaggregated variables, such as regional mean temperature, mean precipitation, and frequency of extremes, that are inferred from up-to-date observational data and more comprehensive climate models.**

> **CONCLUSION 4-3 In the near term, linear pattern scaling, although subject to numerous limitations, provides an acceptable approach to estimating some regionally disaggregated variables from global mean temperature and global mean sea level. If necessary, projections based on pattern scaling can be augmented with high-frequency variability estimated from observational data or from model projections. In the longer term, it would be worthwhile to consider incorporating the dependence of disaggregated variables on spatial patterns of forcing, the temporal evolution of patterns under stable or decreasing forcing, and nonlinearities in the relationship between global mean variables and regional variables.**

UNCERTAINTY PROPAGATION

The climate module will require an uncertainty sampling strategy consistent with the overall strategy for SC-CO_2 uncertainty quantification. Following the uncertainty quantification approach discussed in Chapter 2, the climate module requires two key inputs. First, it requires an emissions projection from the socioeconomic module to drive changes in the Earth system response. Second, it requires a set of parameters to set the response of a simple Earth system model. As discussed above, the joint distributions of key metrics in the model (i.e., ECS, TCR, TCRE, and IPT) will be obtained from IPCC assessments or similar expert assessment processes (see Recommendation 4-2, above). From these distributions, the climate module requires samples of parameters that represent the uncertainty in the model response consistent with current scientific knowledge. These discrete samples could be generated using a large Markov chain Monte Carlo approach (n ~ 100k) or using smaller representative samples, such as Latin hypercube sampling techniques (n ~ 1,000) based on the joint distributions discussed above. At this stage, the simple model would simulate future changes in climate by choosing a single emissions projection and a single set of input parameters from the distributions. This approach would be repeated for each emissions projection produced by the socio-economic module to produce an ensemble of future climate change simulations of global mean surface temperature and CO_2 concentrations.

From the ensemble of model outputs, additional outputs will need to be extracted for the other components of the climate module (sea level rise, pH, and disaggregated variables). Generating probabilistic outputs from these three components requires a similar sampling strategy appropriate for each component to represent uncertainty conditional on the model projection. For sea level rise, the semi-empirical model provides direct estimates of the uncertainty. For pH, the component has very little uncertainty, and each model projection output generates a single pH value (i.e., no cascade of uncertainty).

For the disaggregation of variables, as for sea level rise, a cascade of uncertainty is desirable in the longer term but may not be feasible in the near term. The source for this additional uncertainty has a large set of possibilities. Some standard options are stochastic generating functions (e.g., Fowler et al., 2007), sampling from existing observations (e.g., Wilby et al., 2002), and sampling from available full complexity models (e.g., Schlosser et al., 2012). Developing tools for generating these uncertainty distributions is a substantial longer-term research agenda.

LIMITATIONS OF SIMPLE EARTH SYSTEM MODELS

In complex climate models, the parameters described in Box 4-1 (above)—ECS, TCR, TCRE, and IPT—are resultant behaviors of the climate system, not input parameters. They arise from physical properties of the Earth system, such as the heat capacity of the ocean and the magnitude of different feedbacks that amplify or dampen the temperature change caused by radiative forcing. The strength of these feedbacks depends on the state of the climate; they are not generally constant, and they may vary in response to the magnitude of forcing and spatial pattern of forcing, as well as over time (Knutti and Rugenstein, 2015).

By contrast, in simple Earth system models at least some of these metrics are input parameters. For example, the IWG analysis prescribes values for ECS in DICE, FUND, and PAGE that, along with model-specific parameters, define the relationships between ECS and TCR, TCRE, and IPT. The simple Earth system model described in this chapter is designed such that all four of these metrics may be varied independently. This approach is necessary to accurately capture the joint uncertainty distribution of the metrics, including their co-variation. The committee suggests that, whatever simple Earth system model is used, parameters should be varied so that, at a minimum, the joint distribution of model responses as characterized by these metrics is consistent with up-to-date observational constraints and model-derived knowledge. However, the model would retain the assumption of constant feedbacks that underlie past simple Earth system models. It is therefore important to be aware of three key limitations of this assumption and the use of ECS, TCR, TCRE, and IPT as parameters.

The first limitation is that these metrics are all defined with respect to a reference state, such as the preindustrial state of Earth. They are not, in the real world or in complex climate models, constrained to be constant as they often are in simple models. The feedbacks that control ECS may change. As one example, cloud feedbacks can exhibit state dependence that is represented in more comprehensive models (Crucifix, 2006; Yoshimori et al., 2009; Andrews et al., 2012; Armour et al., 2012; Caballero and Huber, 2013; Bloch-Johnson et al., 2015) but not in simple models that specify a fixed ECS value. As another example, a rising tropopause can lead to an increase in the tropical water-vapor feedback with temperature (Meraner et al., 2013). State-dependent feedbacks can also be related to long-term changes in ocean circulations (e.g., Senior and Mitchell, 2000; Ringer et al., 2006; Yokohata et al., 2008; Caldwell and Bretherton, 2009), land-surface conditions (e.g., Hirota et al., 2011), ocean carbon uptake (e.g., Schwinger et al., 2014), and the cryosphere (e.g., Hakuba et al., 2012).

The second limitation is that these parameters are diagnosed using tests that hold certain elements of the climate system constant. This inactivates certain feedbacks that would change the temperature response to forcing and thus make the parameters a partial representation of the relationship between forcing and warming. As seen in Figure 4-1, this is shown by the exclusion or inclusion of different processes in the boxes defining equilibrium climate, transient climate, and the coupled climate/carbon cycle. As conventionally defined and assessed, ECS includes atmospheric feedbacks (driven by changes in clouds, water vapor concentration, and the lapse rate) and feedbacks involving snow and sea ice cover (Flato et al., 2013). The temperature response to forcing may also involve vegetation, dust, or ice sheet feedbacks. Earth system models may capture some of these additional feedbacks, but simple Earth system models often do not, and they are generally held constant when diagnosing ECS in general circulation models and those of intermediate complexity.

The experiments to assess ECS and TCR prescribe CO_2 concentrations, so carbon cycle feedbacks are also excluded. If these other feedbacks are predominantly positive, then on the timescales on which they are operative, measures such as ECS and TCR will understate the expected warming. As discussed above, the processes in Figure 4-1 shown for the simple Earth system model, which include land and ocean carbon cycle feedbacks, give rise to a CO_2 warming that takes millennia to reverse. Feedbacks affecting albedo or emissivity or adding new net sources of carbon (e.g., carbon dioxide and methane emissions from melting permafrost) would increase the warming response to cumulative emissions beyond that indicated by TCRE.

The third limitation is that three important feedbacks are excluded from ECS and TCR: the response to changes in albedo related to land ice, changes in albedo and transpiration related to land cover changes and the dust/aerosol feedbacks that impact biogeochemical cycles. Geological data suggest that these feedbacks may amplify warming by about 50 percent relative to that expected based on ECS alone (PALAEOSENS Project, 2012). As Earth system models develop further, dynamic vegetation models will partially account for these feedbacks (Ciais et al., 2013; Flato et al., 2013), although the representation of fundamental structure and related feedbacks in the land and ocean carbon cycles remains a developing area (Ciais et al., 2013, Sec. 6.4). Given these limitations, the Earth system models used to investigate metrics such as TCRE may not fully account for the full suite of feedbacks.

Nonetheless, the linear approximations underlying such metrics as ECS, TCR, and TCRE have provided a great source of insight over the past half-century of climate research (National Academy of Sciences, 1979; Hansen et al., 1981, 1984; see also Manabe and Wetherald, 1967) and remain reasonable for use in estimating the global mean temperature response to forcing for purposes of estimating the SC-CO_2. However, they ought to be used with awareness of structural uncertainties that become increasingly important on multicentury timescales.

CONCLUSION 4-4 The linear approximations underlying both simple Earth system models and the metrics ECS, TCR, and TCRE are imperfect. For example, current research suggests it is more likely than not that the warming response to an increase in forcing increases in a warmer global mean climate. Likewise, TCRE may decrease with warming less quickly than indicated by many climate models of intermediate complexity. Nonlinearities may affect both the baseline response of global temperature to forcing and the response of temperature to a pulse emission of CO_2, particularly on centennial and longer timescales. These and other structural uncertainties imply that projections based on simple Earth system models understate long-run climate uncertainty in ways that are difficult to quantify. This uncertainty will affect estimates of the probability distribution of the SC-CO_2, particularly for low discount rates that give significant weight to multicentennial climate responses.

NEEDS FOR FUTURE RESEARCH

This chapter highlights a number of areas in which future Earth system modeling research could improve estimation of the social cost of carbon. Conclusion 4-5 details those areas.

CONCLUSION 4-5 Research focused on improving the representation of the Earth system in the context of coupled climate-economic analyses would improve the reliability of estimates of the SC-CO$_2$. In the near term, research in six areas could yield benefits for SC-CO$_2$ estimation:

- **coordinated research to reduce uncertainty in estimates of the capacity of the land and ocean to absorb and store carbon, especially in the first century after a pulse release, applied to a range of scenarios of future atmospheric composition and temperature;**
- **coordinated Earth system model experiments injecting identical pulses of CO$_2$ and other greenhouse gases in a range of scenarios of future atmospheric composition and temperature;**
- **development of simple, probabilistic sea level rise models that incorporate the emerging science on ice sheet stability and that can be linked to simple Earth system models;**
- **systematic assessments of the dependence of patterns of regional climate change on spatial patterns of forcing, the relationship between regional climate extremes and global mean temperature, the temporal evolution of patterns under conditions of stable or decreasing forcing, and nonlinearities in the relationship between global means and regional variables;**
- **systematic assessments of nonlinear responses to forcing in Earth system models and investigations into evidence for such responses in the geological record; and**
- **the development of simple Earth system models that incorporate nonlinear responses to forcing and assessments of the effects of such nonlinear responses on SC-CO$_2$ estimation.**

In the longer term, more comprehensive climate models could be incorporated into the SC-CO$_2$ estimation framework. However, the major focus of current model research is on increasing resolution and comprehensiveness, rather than on expanding the ability of comprehensive models to be used for risk analysis. SC-CO$_2$ estimation would be advanced by an expanded focus on probabilistic methods that use comprehensive

Earth system models, including the use of comprehensive models to represent low-probability, high-consequence states of the world, as well as the use of decision support science approaches to identify and evaluate key decision-relevant uncertainties in Earth system models.

5

Damages Module

This chapter addresses many of the specific issues raised by the IWG for the committee's consideration and provides suggestions for a path forward. It concludes that, in the longer term, the development of a new damages module, satisfying the scientific criteria stated in Recommendation 2-2, in Chapter 2 (scientific basis, uncertainty characterization, and transparency), and addressing some of the challenges identified by the committee and by the IWG in its 2010 *Technical Support Document*, is merited. Since such a research effort is likely to consume significant resources and time, this chapter also recommends a set of improvements the IWG could undertake in the near term.

The first section below reviews the damage components of the integrated assessment models used to estimate the social cost of carbon (SC-IAMs).[1] The second section discusses alternate approaches to estimating climate damages as well as some of the recent literature on damage estimation. The third section provides the committee's recommendations for improvements in the near term. In the final section the committee offers recommendations for a new damage module that could be developed in the longer term and outlines its properties.

---

[1]These are the three integrated assessment models widely used to produce estimates of the SC-CO_2: the Dynamic Integrated Climate-Economy (DICE) model, the Framework for Uncertainty, Negotiation and Distribution (FUND) model, and the Policy Analysis of the Greenhouse Effect (PAGE) model (see Chapter 1).

CURRENT IMPLEMENTATION OF THE
DAMAGE COMPONENTS IN SC-IAMS

Currently, the damage component of an SC-IAM translates streams of socioeconomic variables (e.g., income and population and gross domestic product [GDP]) and physical climatic variables (e.g., changes in temperature and sea level) into streams of monetized damages over time. To do this, it must represent relationships among physical variables, socioeconomic variables, and damages. To date, the SC-IAMs and related literature consists of damage representations that are either simple and global (e.g., global damages as a function of global mean temperature) or sectorally and regionally disaggregated (e.g., agricultural damages as a function of regional temperature, precipitation change, and CO_2 concentrations).

The damage formulations in the SC-IAMs differ substantially in their sectoral and regional disaggregation of damages, functional forms, drivers of damages, and consideration of parametric uncertainty: see Table 5-1. All three SC-IAM damage components take global mean temperature, global mean sea level, and socioeconomic projections (global population and GDP) as inputs for computing damages. The models differ in their use of the drivers of damages with respect to other climate variables (e.g., CO_2 concentrations, regional temperature), regional socioeconomic projections and sectoral detail (e.g., the agricultural share of the economy, energy efficiency of space cooling and heating), demographic detail (e.g., population density), and other factors. The models also vary in the representation of adaptation, which is implicit in the DICE parameterization, explicit in FUND and PAGE and endogenous only in FUND.

The IWG currently runs each of the SC-IAMs in a simulation mode with information passed from one module to another in a once-through fashion. Thus, the models do not optimize the social response to climate change (except for FUND's adaptation to sea level rise). There are varying degrees of feedbacks to socioeconomic elements (e.g., through effects on GDP or capital stocks) and climate (e.g., through effects on emissions or albedo) in the IWG SC-IAMs (shown in Table 5-1).

In the SC-IAMs, all damages are represented in terms of dollars as fractions of global or regional GDP. Damages therefore scale with the size of the economy, with the rate varying across models and sometimes regions (Rose et al., 2014b). Global damages are a simple summation across sectors and regions (or just across sectors in the case of DICE). Physical units are computed first for some damages, such as mortality and morbidity effects in FUND, but not for all damages in all three models.

The current approach to damage calculations taken by the IWG, to varying degrees, considers three kinds of uncertainty—input (temperature and CO_2 concentration changes, sea level rise, and socioeconomic), parametric (within model), and structural (via the differences in damage

TABLE 5-1 Structural and Implementation Characteristics of Damage Components in SC-IAMs

Characteristic	DICE 2010	FUND 3.8	PAGE 2009
Regions	1 region	16 regions	8 regions
Damage Sectors	2 sectors: sea level rise, aggregate non-sea level rise*	14 sectors: sea level rise, agriculture, forests, heating, cooling, water resources, tropical storms, extratropical storms, biodiversity, cardiovascular and respiratory mortality, vector borne diseases, morbidity, diarrhea, migration	4 sectors: sea level rise, economic, noneconomic (i.e., not in GDP), discontinuity (e.g., abrupt change or catastrophe)
Sea Level Rise Damage Specification (Fraction of Income)	Quadratic function of global sea level rise	Additive functions for coastal protection costs, dryland loss, and wetland loss, based on an internal cost-benefit rule for optimal adaptation	Power function of global sea level rise
Drivers of Sea Level Rise Damage	Global mean sea level rise, income	Global mean sea level rise, dryland value, wetland value, topography, protection cost, population density, income density, per capita income	Global mean sea level rise, regional coast length scaling factor relative to European Union, adaptation capacity and costs, per capita income, income
Non-Sea Level Rise Damage Specifications (Fraction of Income)	Quadratic function of global temperature	Uniquely formulated nonlinear functions by sector (see Anthoff and Tol, 2014)	Power function of regional temperature

continued

TABLE 5-1 Continued

Characteristic	DICE 2010	FUND 3.8	PAGE 2009
Non-Sea Level Rise Damage Drivers	Global mean temperature, income	Global mean temperature, CO_2 concentrations (for carbon fertilization and storms), population, income, technological change	Regional temperature, regional scaling factor relative to the European Union, adaptation capacity and costs, population, income
Adaptation	Implicit (damages net of adaptation)	Explicit for agriculture and sea level rise, implicit otherwise (econometric studies of net response to warming)	Two types of exogenous fixed adaptation policy that reduce impacts for a cost
Climate Benefits	Implicit (damages net of benefits)	Explicit outcome of certain sectoral damage functions (e.g., avoided heating demand, agriculture benefits from CO_2 fertilization)	Assumes small economic benefits at low levels of warming
Damages Due to Abrupt Climate Change	Included in calibration of aggregate damages not from sea level rise*	No explicit representation	Unspecified "discontinuity" impact occurs with a positive probability at global average temperature changes greater than 3 °C
Feedbacks from Damages	Damages affect global income, which affects future global capital stocks and income levels, but projected emissions are unaffected	No economic feedback	No economic feedback

*These damages are an aggregate based on a calibration of sectoral damages according to Nordhaus and Boyer (2000) and rescaled using external aggregate damage information. For additional details, see note [a] to Table 5-2.
SOURCE: Adapted from Rose et al. (2014b, Table 6-1).

formulations among the three models). Input uncertainty is considered in the form of alternative climate and socioeconomic input projections (see Chapters 3 and 4). Parametric uncertainty is considered in the damage formulations of two of the current SC-IAMs, FUND and PAGE. DICE in its standard formulation, used by the IWG, does not consider parametric uncertainty, although a variety of studies have explored some forms of parametric or structural uncertainty with versions of the DICE damage function (e.g., Nordhaus and Popp, 1997; Azar and Lindgren, 2003; Keller et al., 2004; Ackerman et al., 2010; Kopp et al., 2012; Cai et al., 2016; Lemoine and Traeger, 2016). The parametric uncertainty specifications in FUND and PAGE differ, with FUND representing larger uncertainty in annual damages through 2100, but less than PAGE after 2100, and PAGE exhibiting higher average annual damages (Rose et al., 2014b). Structural uncertainty is considered to a degree in the IWG's framework by including the three SC-IAMs. However, as Table 5-2 shows, the most recent SC-IAM formulations for PAGE and DICE exhibit some degree of dependency on the other models (see discussion below on model dependency).

Another attribute of the SC-IAMs that underpin the current IWG estimates is that much of the research on which they are based is dated. As Table 5-2 shows, the damage formulations do not in many cases reflect recent advances in the scientific literature (e.g., some using sources not more recent than the 1990s and early 2000s).

Figure 5-1 illustrates that there are significant differences across models in global damage response to key input drivers of damages. DICE and PAGE yield higher damages for a given level of warming and income and

TABLE 5-2 Literature Sources for Current SC-IAM Damage Component Specifications

Model (Version)	Damage Type	Study	Basis for Damage Estimate
DICE 2010[a]	Aggregate non-sea level rise	Literature surveys Intergovernmental Panel on Climate Change (2007a) and Tol (2009) used to rescale Nordhaus and Boyer (2000) sectoral damages[b]	Calibration
	SLR coastal impacts	Undocumented	

continued

TABLE 5-2 Continued

Model (Version)	Damage Type	Study	Basis for Damage Estimate
FUND 3.8	Agriculture	Kane et al. (1992), Reilly et al. (1994), Morita et al. (1994), Fischer et al. (1996), Tsigas et al. (1996)	Calibration
		Tol (2002b)	Income elasticity
	Forestry	Perez-Garcia et al. (1995), Sohngen et al. (2001)	Calibration
		Tol (2002b)	Income elasticity
	Energy	Downing et al. (1995, 1996)	Calibration
		Hodgson and Miller (1995)	Income elasticity
	Water resources	Downing et al. (1995, 1996)	Calibration
		Downing et al. (1995, 1996)	Income elasticity
	Coastal impacts	Hoozemans et al. (1993), Bijlsma et al. (1996), Leatherman and Nicholls (1995), Nicholls and Leatherman (1995), Brander et al. (2006)	Calibration
	Diarrhea	Global Burden of Disease 2000 estimates[c]	Calibration
		Global Burden of Disease 2000 estimates[c]	Income elasticity
	Vector-borne diseases	Martin and Lefebvre (1995), Martens et al. (1995, 1997), Morita et al. (1994)	Calibration
		Link and Tol (2004)	Income elasticity
	Cardiovascular and respiratory mortality	Martens (1998)	Calibration
	Storms	CRED EM-DAT database,[d] World Meteorological Organization (2006)	Calibration
		Toya and Skidmore (2007)	Income elasticity
	Ecosystems	Pearce and Moran (1994), Tol (2002a)	Calibration

TABLE 5-2 Continued

Model (Version)	Damage Type	Study	Basis for Damage Estimate
PAGE09	SLR	Anthoff et al. (2006)[e]	Calibration and income elasticity
	Economic	Warren et al. (2006)[f]	Calibration
	Noneconomic	Warren et al. (2006)	Calibration
	Discontinuity	Lenton et al. (2008), Nichols et al. (2008), Anthoff et al. (2006), Nordhaus (1994a)[g]	Calibration
	Adaptation costs	Parry et al. (2009)	Calibration

[a]The committee assembled the following information related to the calibration of DICE 2010 based on communications with William Nordhaus, Nordhaus and Boyer (2000), and Nordhaus (2010). DICE global damages have historically been calibrated to the aggregate results of another model, RICE, which has regional and sectoral damage calibrations. RICE 2000 is the last full set of regional and sectoral damage estimates that are fully documented for the DICE/RICE family of models, and DICE 2000's global estimate was calibrated to RICE 2000. These estimates were based on Nordhaus and Boyer (2000) and calibrated at the sector level using the following sources as the main references: agriculture (Darwin et al., 1995), health (Murray et al., 1996), energy (Nordhaus and Boyer, 2000), recreation (Nordhaus and Boyer, 2000), human settlements and natural ecosystems (Nordhaus and Boyer, 2000), coastal impacts (Yohe and Schlesinger, 1998), and catastrophic damages (Nordhaus, 1994a; Nordhaus and Boyer, 2000). Updates to DICE/RICE prior to DICE 2013 have used the same sectoral breakdown of damages as RICE 2000 but changed the aggregate based on further information. According to Nordhaus (2010), Tol (2009), and Intergovernmental Panel on Climate Change (2007a) were the additional information used for DICE/RICE 2010. However, the specifics of the recalibration are not available. For information regarding DICE 2007, which was used for the Interagency Working Group on the Social Cost of Carbon (2010) SCC estimates, see Nordhaus (2007, 2008).

[b]Tol (2009) is a survey of global damage studies, some of which report impacts estimated by earlier versions of the SC-CO$_2$ models.

[c]See http://www.who.int/healthinfo/global_burden_disease/estimates_regional_2000/en [November 2016].

[d]See http://www.emdat.be [November 2016].

[e]Anthoff et al. (2006) is a study of coastal impacts that uses an earlier version of FUND 2.8.

[f]Warren et al. (2006) is a review of damage modeling in earlier versions of four integrated assessment models: DICE/RICE 1999, MERGE 1995 and 2004, PAGE2002, and FUND 2.9.

[g]Nordhaus (1994a) is an expert elicitation on climate catastrophes, and is also used as the basis for catastrophic impacts in DICE prior to 2013.

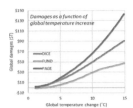 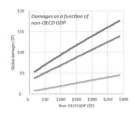 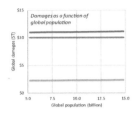

FIGURE 5-1 Annual global damages ($2005 per year) in 2100 by model as a function of major individual drivers of damages holding other inputs constant.
NOTES: The figures were developed running only the damage component of each SC-IAM with standardized input assumptions and central parameter values. The individual figures were developed by varying the relevant single driver and fixing other drivers. When fixed, drivers had the following global values in 2100—GDP $370 trillion, population 9.6 billion, mean temperature increase 4 °C. Y-axis ranges vary. This sensitivity analysis misses some cumulative damage effects over time (e.g., in DICE, sea level rise and reductions in capital stock and GDP). The temperature domain in the first figure is consistent with the range of outcomes resulting from probabilistic analyses of the FUND and PAGE climate component. See Rose et al. (2014b) for probabilistic climate and damage results, as well as damages relative to Organisation for Economic Co-operation and Development (OECD) income. The SC-IAM damage components respond differently to changes in richer versus poorer country incomes.
SOURCE: Developed from Rose et al. (2014b, Chapter 6).

are much more responsive to both temperature change and income than FUND, while none of the models are particularly responsive to global population size.[2] Communicating and providing scientific justification for these differences is critical, as discussed below. For each model, the slope of the temperature response is indicative of the projected incremental damages resulting from a pulse of CO_2.

Figure 5-2 displays the estimated incremental damages over time produced by the three SC-CO_2 models in response to an identical incremental change in projected temperature from a CO_2 emissions pulse in

---

[2]The population results in Figure 5-1 are from a sensitivity analysis that scales global population with the regional distribution fixed. Population enters each of the SC-IAMs differently. In DICE, population affects total factor productivity, income, and the capital stock. In FUND, population affects per capita income and is an explicit input variable in a number of individual damage categories (water resources, energy consumption, ecosystems, various human health damage categories, tropical storms). In PAGE, population affects per capita income, which enters into each category of damages within the model.

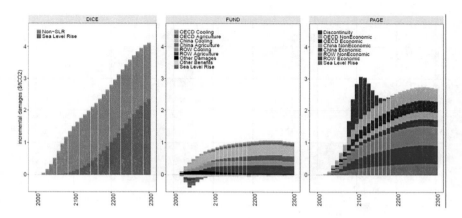

FIGURE 5-2 Incremental annual damages ($2005/tCO$_2$ per year) to 2300 and key factors for the SC-IAM damage components with standardized climate and socioeconomic inputs.

NOTES: Shown are incremental annual damages from standardized projected increases in annual temperature and CO$_2$ concentrations. The standardized incremental climate projections were derived from adding a CO$_2$ emission of 1 GtC (3.7 Gt CO$_2$) in 2020 to the IWG's highest emissions (and corresponding socioeconomic) scenario.

SOURCE: Developed from Rose et al. (2014b, Figures 6-40 and 6-41).

2020.[3] Annual incremental damages differ in sign, magnitude, timing, and regional and sectoral composition. Underlying Figure 5-2 are significant differences in total global damage levels. For example, DICE and PAGE produce annual global damages in 2100 that are four times larger than those from FUND for the same reference climate and socioeconomic future used for Figure 5-2. The models also differ notably in the size and sensitivity of their responses to key uncertain inputs, as shown in Figure 5-1 (Rose et al., 2014b).

The differences in model characteristics (shown in Table 5-1) drive the differences in results, with specific characteristics playing a prominent

---

[3]Each model's damage component was first run with identical reference temperature, CO$_2$ concentration, and socioeconomic projections using the IWG's highest emissions (and corresponding socioeconomic) scenario, and then run again with identical incrementally higher temperature and CO$_2$ concentration projections resulting from a 2020 1 billion ton carbon (3.7 Gt CO$_2$) emissions pulse. Projected incremental damages over time are the difference between the projected damages in the two scenarios. The results in Figure 5-1 reflect differences across models in the modeling of sea level, regional temperatures, and damages, which are all driven by global average temperature change in the models (see Table 5-1). See Rose et al. (2014b) for more details and discussion.

role. For example, in DICE, damages are based on quadratic functions of temperature and sea level rise; in FUND, net benefits in the agricultural sector result at lower warming levels, adaptation addresses much of the risk from sea level rise, cooling energy demand costs are a large fraction of damages, and "catastrophic" damages are not included; and, in PAGE, regional damages are computed by scaling damages between regions, and a large fraction of damages are from those that do not directly impact GDP and an unspecified discontinuity damage.

The committee evaluated the damage components of the three SC-CO$_2$ IAMs according to the criteria in Recommendation 2-2, in Chapter 2. Overall, the damage formulations of the three models used by the IWG, and the overall IWG damage modeling approach, differ in the extent to which they satisfy the three criteria: scientific basis, uncertainty characterization, and transparency. None of the damage components fully satisfies all the criteria. Analysis of the IWG documentation, individual model documentation, and outside research suggest a number of elements in the current damage functions, individual model results, and damage components as a whole that can be improved.

The committee notes that the Interagency Working Group on the Social Cost of Carbon (2010) identified a number of potential shortcomings and critiques of the current damage formulations, which are discussed further below. These include:

- incomplete treatment of noncatastrophic damages;
- incomplete treatment of potential catastrophic damages;
- uncertainty in extrapolation of damages to high temperatures;
- incomplete treatment of adaptation and technological change;
- omission of risk aversion with respect to high-impact damages;
- failure to incorporate intersectoral and interregional interactions; and
- imperfect substitutability of consumption for environmental amenities.

CURRENT LITERATURE ON CLIMATE DAMAGES

The committee defines *climate impacts* as the biophysical or social effects driven by climate change (e.g., changes in land productivity, mortality, morbidity, water supply, coastal flooding, or conflict) and *climate damages* as the monetized estimates of the social welfare effects of climate impacts (see Box 2-2 in Chapter 2). Impacts estimates are either an explicit input or implicit element of projected damages.

The damage component of a reduced-form IAM is composed of damage functions that monetize climate change effects, with functional forms and calibrations that are in some way derived from and calibrated to more detailed climate change impacts and damage analyses, other parameters (e.g., economic elasticities), and a modeler's judgment (see Table 5-2). SC-IAM damage functions are thus constrained by the available literature, and they typically need to extrapolate beyond the relationships characterized in the detailed supporting analyses, for instance, beyond the warming levels evaluated or locations studied.

The scientific literature has produced studies of damages and impacts using physical process models, structural economic models, and empirical models.

- Physical process models describe the dynamics of a physical process to identify a climate change–induced physical impact and evaluate its implications: for example, crop models assess the impact of temperature, precipitation, CO_2 concentrations, and other drivers on plant productivity.
- Structural economic models describe the structure and dynamics of economic decisions and markets to evaluate the net economic implications of climate-induced physical changes: for example, they can assess the economic consequences of climate-related changes in the productivity of land or the labor force, as well as the demand for heating or cooling.
- Empirical models estimate statistical relationships between weather (short-run) or climatic (long-run) variables and human or ecological responses from historical data: for example, they are used to estimate dose-response functions between exposure to temperature and mortality.[4]

The literature includes impact and damage research that varies in scope. It includes studies of individual and multiple sectors, studies at local, national, and global geographic scales, studies using higher and lower spatial resolution, studies that model different processes and interactions, and studies that focus on market and nonmarket damages. The differences in methodologies and in scope create challenges for users trying to synthesize understanding of impacts or damages. For instance, structural and empirical methods are fundamentally different from one

---

[4]Early empirical work largely relied on cross-sectional techniques (i.e., comparing the relationship between climate and outcomes across space, potentially capturing other factors that lead to spatial variability). The most recent empirical literature has employed methodological insights from the causal inference literature, which has resulted in numerous estimates for a number of sectors that actually reflect causal relationships between weather/climate and economically relevant variables, represent populations of interest in damage function calibration, incorporate nonlinearities, and empirically reflect historical forms of adaptation.

another and, as a consequence, they produce results that may not be directly comparable. These comparability and scope issues need to be addressed in some way when developing damage functions.

As discussed above, each of the current SC-IAM damage components has some direct or indirect link to the damages literature of the 1990s and early 2000s. The literature has, however, evolved substantially since then. This more recent literature yields economic estimates that could be integrated into SC-CO$_2$ modeling in the near term. The research community has also initiated activities that will yield useful impacts and damages information in the longer term. These activities are important to monitor and are discussed in the context of our proposals for the longer term, in the final section of this chapter.

Table 5-3 lists a number of studies that could be used as resources for a near-term update to individual SC-IAM damage formulations and the damages module as a whole. This table is not comprehensive, and this

TABLE 5-3 Selected List of Climate Damages Literature for a Near-Term Update

Impacts	Regions	References and Sources of Information
Health, infrastructure, electricity, water resources, agriculture and forestry, ecosystems	United States	Waldhoff et al. (2015), Marten et al. (2013) www2.epa.gov/cira [December 2016]
Agriculture, energy, river floods, forest fires, transport infrastructure, coastal areas, tourism, human health, habitat suitability	Europe	Ciscar et al. (2011, 2014) https://ec.europa.eu/jrc/en/peseta [January 2017]
Agriculture, labor productivity, mortality, property and violent crime, energy demand, coastal storms and inundation	United States	Houser et al. (2015)
Heat extremes and health, agriculture and land use, tropical cyclones, sea level rise, drought and conflict	Global	https://chsp.ucar.edu/brace [December 2016]
Sea level rise, agricultural productivity, heat effects on labor productivity, human health, tourism flows and households' energy demand	Global	Roson and Sartori (2010, 2016)

TABLE 5-3 Continued

Impacts	Regions	References and Sources of Information
Sea level, agriculture, and energy demand	Global	Bosello et al. (2012)
Agriculture	Global	Reilly et al. (2007), Kyle et al. (2014), Nelson et al. (2014)
Coastal damages	Global	Diaz (2016)
Energy demand	Global	Isaac and van Vuuren (2009), Mima and Criqui (2009), Labriet et al. (2013), Zhou et al. (2013)
Energy supply	Global	Mima and Criqui (2009), Labriet et al. (2013), Kyle et al. (2014)
Water	Global	Hanasaki et al. (2013), Hejazi et al. (2014), Schlosser et al. (2014), Kim et al. (2016)
Ecosystem services	Global	http://www.naturalcapitalproject.org/invest [December 2016]
Empirical adaptation response	Regional, multiple sectors (agriculture, energy, mortality)	Auffhammer and Aroonruengsawat (2012), Barreca et al. (2015), Hsiang and Narita (2012); Hsiang and Jina (2014), Butler and Huybers (2013)

section does not review and assess the literature; the time frame for this report did not allow for such an activity. This newer literature needs to be considered and, to the extent possible, incorporated in the near-term update.

Since the studies that are used to calibrate the SC-IAMs were conducted, there has been significant progress in research into both market and nonmarket damages, and in methods using both empirical and structural models. In the future, the calibration of damage functions needs to be compared to point estimates from newer literature as either validation of or justification for updates; and, where possible, assessment of the damage calibration using hindcasting and comparisons to empirical studies will be valuable. Going forward, there are necessary and complementary roles for both empirical and structural modeling.[5]

NEAR-TERM IMPROVEMENTS IN
SC-CO$_2$ DAMAGE ESTIMATION

In the near term, the IWG has two options for developing the damages module of an integrated SC-CO$_2$ estimation framework: an improved damage component of a single SC-IAM (or another reduced-form IAM) or improved damage components from multiple SC-IAMs (or other reduced-form IAMs). While the committee does not recommend a specific path for the IWG, it recommends a set of steps for *any* damage component and module used in a near-term update of the SC-CO$_2$ estimates.

> **RECOMMENDATION 5-1 In the near term, the Interagency Working Group should develop a damages module using elements from the current SC-IAM damage components and scientific literature. The damages module should meet the committee's overall criteria for scientific basis, transparency, and uncertainty characterization (see Recommendation 2-2, in Chapter 2) and include the following four additional improvements:**
>
> 1. **Individual sectoral damage functions should be updated as feasible.**
> 2. **Damage function calibrations should be transparently and quantitatively characterized.**
> 3. **If multiple damage formulations are used, they should recognize any correlations between formulations.**
> 4. **A summary should be provided of disaggregated (incremental and total) damage projections underlying SC-CO$_2$ calculations, including how they scale with temperature, income, and population.**

These improvements are discussed in the four sections below.

In the near term, the IWG will need to choose which damage formulations to include in the damages module. Whether the IWG includes multiple formulations or only a single one, the damage formulations need to be consistent with the recent literature. The IWG's choice in this matter has implications for the level of disaggregation required from

---

[5]Information obtained through a focused literature review performed for the committee by Frances Moore (University of California, Davis) and Delavane Diaz (Electric Power Research Institute (EPRI).

the socioeconomic and climate modules in the near term. It is important to differentiate between the spatial and temporal level of aggregation in input data used in calibration of the damage formulation(s) and the level of aggregation represented in an SC-IAM. Calibration of damage formulations may be done using data at a higher resolution than represented in the IAM. The two previous chapters offer guidance on how disaggregation across regions (and sectors) could be accomplished in the near term; early coordination of disaggregation choices in the damages module with the socioeconomic and climate modules will be important for smooth implementation of the committee's recommended modular approach. However, there is no ideal disaggregation level, as there are many factors to consider and tradeoffs with high and low resolution. (See the disaggregation section below for additional discussion.) In addition, documentation for each damage formulation—its implementation (i.e., how it is run and how uncertainty is modeled), and aggregation across formulations—needs to be provided with sufficient detail and justification for the scientific community to understand and assess the modeling.

Below, guidance is provided for a near-term revision by discussing each of the four points in Recommendation 5-1 above. In addition, Appendix G presents model-specific improvements for each of the SC-IAM damage formulations that could be pursued during a near-term update if the IWG wished to continue with some elements of one or more of the SC-IAMs. The IWG may also wish to consider additional damage formulations that have been published in the peer-reviewed literature (e.g., Roson and van der Mensbrugghe, 2012). Any alternative formulations, their implementation, and potential multi-model integration would also need to be evaluated applying the criteria in Recommendations 2-2 (in Chapter 2) and 5-1 (above).

Updating Individual Sectoral Damage Functions

As discussed above, research on climate damages has advanced beyond the studies underlying the current SC-IAM damage components. A newer and substantial body of additional empirical and structural modelling literature is now available. The literature on agriculture, mortality, coastal damages, and energy demand provide immediate opportunities to update the SC-IAMs. For example, Moore et al. (2016) provide a possible blueprint for how to achieve this for FUND. Points of departure in terms of resources that could be used for updating damage components include the studies listed in Table 5-3 (above), the empirical studies reviewed in sources, such as Dell et al. (2014) and Carleton and Hsiang (2016), and other individual peer-reviewed papers with economic damage estimates (based on either structural economic models or empirical estimates).

A key challenge as noted above will be to determine how to use economic damage results from different methods that are not fully comparable. Although many studies do not follow the causal chain all the way to monetized welfare losses and are not global in extent, they still may be used for assessing the calibration of biophysical impacts and damages in particular regions. The comparisons need to be conducted with awareness of the different ways in which the studies account for adaptation. There have been significant improvements in understanding and measurement of adaptive responses for some sectors in empirical and structural modeling, which could be considered in some way in a near-term update. Table 5-3 illustrates that agriculture, energy, mortality, and coastal damages provide some of the most immediate opportunities for updates, with both empirical (last row of Table 5-3) and structural modeling analyses (various rows).

Damage Function Calibrations

The damage formulations currently used in the SC-IAMs are not clearly and adequately justified with regard to how they are parameterized and calibrated and how particular sectors and regions contribute to the overall results. This inadequacy stems from the incomplete documentation of the individual SC-IAMs. DICE and FUND do provide some documentation for the parameterization and calibration of their models, but the accounting of how sectors and regions contribute to the damage function is not transparent. It is not possible to understand with great confidence the actual damage function calibrations and the magnitude of the sectoral contributions, even after investigating different versions of the model code, documentation, and related papers. In addition, PAGE does not provide a detailed description and scientific justification of how its damage component is parameterized.

Going forward, any damage component used in the calculation of the SC-CO_2 needs to provide a clear accounting of the calibration of the damage functions. Such documentation will significantly improve scientific rationale and transparency and allow for improved scientific assessment. For DICE 2010, for example, adequate documentation would mean a clear description of the calibration of the global sea level rise and non-sea-level rise damage functions, as well as details regarding any underlying calibrations at the sector and regional levels. For FUND and PAGE, adequate documentation would entail a clear description of the calibration of the region-sector damage functions. This description will likely require input from the modelers themselves. If the damage functions are updated as detailed in the preceding section, the calibration of these updated functions would need to be documented.

Combining Multiple Damage Formulations

The IWG has pooled the results of three SC-IAMs to estimate the SC-CO_2. Pooling results of multiple SC-IAMs is a method to incorporate structural uncertainty, as each model provides an alternative representation of how damages depend on climate change and other factors. However, when aggregating across models, it is important to consider the degree of dependence of the estimates across models: see Box 5-1. If the models are independent, aggregation of the results provides more information than any single model, but if the models are dependent, combining results may provide little additional information. Moreover, analysts might mistakenly underestimate the degree of uncertainty about the SC-CO_2 if they combine results of dependent models on the assumption that the models are independent.

If the extent of dependence among the models is known, one can estimate the extent to which the structural uncertainty that is captured is reduced, in comparison with a case in which the models are independent. Specifically, one can estimate the number of independent models that would yield an output distribution with a similar spread (Clemen and Winkler, 1985). It is difficult, however, to appropriately characterize the dependence among models. Damage components of all of the SC-IAMs draw on a common literature, yet they use very different functional forms, which contribute to the differences in damages displayed in Figure 5-2 (above). In addition, some of the damage components draw on results of the damage components of current or previous versions of the SC-IAMs (see Table 5-2). The use of a common literature is appropriate; it is desirable that models be based on the best available scientific evidence, and a model that ignored relevant parts of the literature could be improved by including those parts. The reliance on damage components of other SC-IAMs is more problematic. This reliance induces dependence among the models that affects the extent to which structural uncertainty is captured by using multiple models. This dependence needs to be recognized when aggregating the model outputs, but it is not clear how to characterize the dependence and quantify its effect on the representation of structural uncertainty.

Whether the models are independent or not does not affect the interpretation of the central value of the distribution of SC-CO_2 estimates obtained by pooling results across the models. If each of the models is judged to be unbiased (in the statistical sense of not systematically overestimating or underestimating damages), then each model provides an unbiased estimate of damages. In this case, the average of their results is also unbiased. The degree of independence affects the spread of the results but not the central value.

BOX 5-1
Model Dependence and Structural Uncertainty

The IWG characterized uncertainty about the SC-CO$_2$ estimates by producing a frequency distribution using Monte Carlo analysis of each of the three SC-IAMs and then aggregating across the models using equal weights. Specifically, for each of the three discount rates it considered, the IWG produced a frequency distribution of 50,000 realizations from each model (stratifying across the five socioeconomic scenarios and drawing randomly from the probability distributions for equilibrium climate sensitivity [ECS] and model-specific distributions for other parameters) and then pooled these realizations, yielding a frequency distribution of 150,000 realizations.

To understand how lack of independence among models affects the representation of structural uncertainty, consider each realization in the Monte Carlo analysis using a single SC-IAM as an estimate of the SC-CO$_2$ equal to the true value plus an error term. The multiple realizations of a model obtained by taking random draws from the probability distributions of ECS and other inputs yields a distribution of SC-CO$_2$ estimates, conditional on the model structure and the values of the input parameters that are held constant. The mean of this distribution provides a central estimate of the SC-CO$_2$, and the variance provides an estimate of uncertainty. If the model-specific error distribution has a mean of zero, the mean of the realized estimates is by definition an unbiased estimate of the SC-CO$_2$.

Similarly, the frequency distributions obtained by Monte Carlo analyses of the other models yield model-specific distributions of the estimated SC-CO$_2$. If the error distributions of the other models have means of zero, these models also provide unbiased estimates. If the errors are independent across models, then, conditional on inputs that are held fixed across models (e.g., ECS and socioeconomic scenario), the estimates of SC-CO$_2$ from the different models are independent. Pooling estimates from multiple models yields multiple estimates that differ at least in part because of structural uncertainty that is represented by the alternative models.

If the model-specific errors of two or more models are positively correlated, however, pooling estimates across these models yields less variation in the estimates than if the errors are independent. In the extreme case, if the model-specific errors of all the models were perfectly correlated, then pooling their estimates would yield the same distribution as would the use of any one of the models alone. If the model-specific error distributions all have means of zero, the resulting estimates remain unbiased, but lack of independence among the models implies that the distribution obtained by pooling model results captures less structural uncertainty than if the models were independent.

Disaggregated Summaries of Incremental and Total Damage Projections

Going forward, the IWG needs to make intermediate and disaggregated damage projections for both incremental and total damages available. This would include model-specific undiscounted damages over time, regions, and sectors, as well as a characterization of the uncertainty in results. This will improve the transparency and credibility of the individual damage formulations. Given the large potential volume of data, the IWG could provide a representative, summary characterization of the disaggregated damages underlying the SC-CO$_2$ estimates. In addition, the IWG could provide the dataset of intermediate and disaggregated results to the public. See Rose et al. (2014b) for the kind of results the committee suggests be provided in the near term. Two examples are displayed in Figures 5-1 and 5-2.

A DAMAGES MODULE FOR THE LONGER TERM

This section offers a set of desirable characteristics of a damages module that the committee believes can be developed in the longer term, given current scientific understanding. The committee believes that work on such a module could commence immediately and proceed in parallel with implementation of the committee's near-term recommendation, discussed in the preceding Section.

> **RECOMMENDATION 5-2** In the longer term, the Interagency Working Group should develop a damages module that meets the overall criteria for scientific basis, transparency, and uncertainty characterization (see Recommendation 2-2, in Chapter 2) and has the following five features:
>
> 1. It should disaggregate market and nonmarket climate damages by region and sector, with results that are presented in both monetary and natural units and that are consistent with empirical and structural economic studies of sectoral impacts and damages.
> 2. It should include representation of important interactions and spillovers among regions and sectors, as well as feedbacks to other modules.
> 3. It should explicitly recognize and consider damages that affect welfare either directly or through changes to consumption, capital stocks (physical, human, natural), or through other channels.
> 4. It should include representation of adaptation to climate change and the costs of adaptation.
> 5. It should include representation of nongradual damages, such as those associated with critical climatic or socioeconomic thresholds.

Developing a damages module with these characteristics would represent a major advance in understanding the monetary impacts of climate change. In the rest of this section the committee discusses in more detail each of the five features.

Disaggregation of Climate Damages by Region and Sector

Regional and sectoral damage resolution is needed for transparency and to connect estimates to the literature on impacts and damages. However, a priori, there is no ideal disaggregation level. There are a number of factors to consider in determining an appropriate level of disaggregation, including the timescale over which damages are projected and whether the disaggregation is needed for the implementation or calibration of a damages module. In many cases, the level of disaggregation will be determined by the findings available from the literature on impacts and damages and the resolution of economic statistics, computational constraints, and the possible tradeoffs between capturing heterogeneity in climate risks (due to differences in markets, technology, policies, cultures, and physical systems) and feedbacks between affected groups and locations. In addition, the SC-CO$_2$ context matters. For instance, SC-CO$_2$ modeling does not need to have the same spatial and temporal resolution as desired for adaptation planning by a local (e.g., city) decision maker as it would for a national-level decision maker.

Damages could be incorporated in an IAM in one of three ways: (1) using a global reduced-form damages module that is calibrated to spatially and sectorally disaggregated damage formulations, (2) using a damages module that includes spatially and sectorally explicit modeling of relevant processes, or (3) using a directly calibrated and estimated global damages module. DICE 2007 and earlier versions took the first approach, attempting to calibrate a global damage function based on regional and sectoral damage functions that were calibrated to sectoral studies and a reinterpretation of expert elicitation results regarding the possibility of climate-linked economic "catastrophes" (Nordhaus, 1994a; Nordhaus and Boyer, 2000). FUND takes the second approach, with individual reduced-form damage functions for a range of sectors and impact types: agriculture, forestry, water resources and energy consumption, costs of protection against sea level rise, willingness to pay to avoid ecosystem loss, diarrhea, vector-borne diseases, cardiovascular disease, and tropical and extratropical storm damage (Anthoff and Tol, 2014). Though some more complex IAMs incorporate detailed representations of specific damage pathways (e.g., for energy demand), no IAM attempts to be both detailed and comprehensive (Nordhaus, 2014). The social cost of carbon (SCC) took the third approach: it attempted to estimate a total global damage function directly, without a disaggregated calibration. It was based on an interpretation of a meta-analysis of past global damage estimates (Tol, 2009).

A total-damage approach might also be taken based on structured expert elicitation (Nordhaus, 1994a; Pindyck, 2015; Howard and Sylvan, 2016). However, the committee does not recommend an approach based on top-down estimation of a total global damage function because it lacks traceability to damage pathways, may not have a strong scientific rationale, or may not address nonmarket damages (e.g., Dell et al., 2012; Burke et al., 2016). More specific peer-reviewed structured expert elicitation studies that address hard-to-quantify damage categories may be useful in helping to calibrate a damage function to quantitative studies that examine specific impacts.

Structural economic and empirical models, such as those listed in Table 5-3, provide the main resource for calibrating damage formulations. Due to the detailed representation of the weather and climate links to impacts, using either structural economic or empirical models to project future changes requires a high level of spatial and temporal detail in climate and, possibly, in socioeconomic projections, comparable to the level of detail in the past observations with which they are being compared. This level of detail need not necessarily be provided by the climate module of a SC-IAM; however, results from detailed structural economic or empirical models could be used to calibrate relatively simple reduced-form models that require only relatively coarse spatial and temporal detail (as is the case in the current SC-IAMs).

Climate damages do not arise directly from physical climate variables, such as temperature or precipitation. They arise through biophysical or social pathways: agricultural damages arise because temperature and precipitation influence crop yields; labor productivity damages arise because temperatures and humidity affect the quantity and quality of work; and health and longevity are lost because of changes in heat stress and disease. Some physical climate impacts are of potentially great socioeconomic importance, but challenging to translate into dollars: for example, changes in the risk of civil conflict, human migration, or global biodiversity.

Climate damages can occur through a variety of pathways, some quantifiable, some identifiable but hard to quantify, and some unknown. In principle, the SC-CO_2 estimates are intended to represent total economic damages, and thus they are the aggregate over all three types of pathways. However, these types of pathways are successively more difficult to estimate.

In order to provide a satisfactory degree of transparency, it is desirable for the damages module to report impacts in physical units when possible, such as crop yield changes, mortality, or species effects. These natural-unit measures are more straightforward to compare to the impact literature and require fewer intermediary assumptions to estimate than their monetized counterparts. Moreover, reporting physical units for impacts that cannot be monetized allows for their inclusion in regulatory impact analyses, which is consistent with regulatory guidance.[6]

Representation of Important Interactions and Spillovers among Regions and Sectors

Most of the structural and empirical studies that can be used to calibrate a damage function focus on a single type of impact or on the direct effect of climate change on regions in isolation. There is an emerging literature that also incorporates interactions among regions and impacts (e.g., Reilly et al., 2007; Warren, 2011; Diffenbaugh et al., 2012; Taheripour et al., 2013; Baldos and Hertel, 2014; Grogan et al., 2015; Harrison et al., 2016; Zaveri et al., 2016). For example, given global markets, migration, and other factors, effects of a crop failure in India will also have impacts in other countries, and reductions in water availability in one region will have impacts across many regions and sectors.

One set of interactions occurs through market mechanisms, such as trade. For example, the economic impacts of climate change on crop yield in one region will depend in part on the changes in crop yields in other regions. These interactions can be captured by multisectoral, multiregional economic computable general equilibrium (CGE) models. Models of global agriculture and forestry impacts have been developed over more than two decades (e.g., Reilly et al., 1994; Sohngen et al., 2001; Reilly et al., 2007; Roson and van der Mensbrugghe, 2012; Nelson et al., 2014).

Impacts can also interact with each other, and with mitigation policy, through their effects on competition for resources, such as water and land. The relationship between temperature exposure and crop yields depends strongly on whether crops are irrigated (Schlenker and Roberts, 2009; Houser et al., 2015); the ability to irrigate will in turn depend on impacts on water resources.

Some impacts may partially represent adaptations to other impacts; care needs to be taken to avoid double counting. For example, increased demand for space cooling is the major driver of the increased energy costs associated with higher temperatures (e.g., Auffhammer and Mansur, 2014). Yet the widespread adoption of air conditioning significantly reduces the effect of temperature on mortality (Barreca et al., 2013) this paper makes

---

[6]For example, OMB Circular A-4 notes that "Even when a benefit or cost cannot be expressed in monetary units, [an agency] should still try to measure it in terms of its physical units."

two primary discoveries. The mortality effect of an extremely hot day declined by about 80 percent between 1900-1959 and 1960-2004. As a consequence, days with temperatures exceeding 90 °F were responsible for about 600 premature fatalities annually in the 1960-2004 period, compared to the approximately 3,600 premature fatalities that would have occurred if the temperature-mortality relationship from before 1960 still prevailed. Similarly, the sensitivity of labor supply to temperature depends to a large extent on whether workers are protected from outdoor temperatures (Graff Zivin and Neidell, 2014). Thus, increases in the impact of energy demand impact may be offset by decreases in other impacts.

In the SC-IAMs, damages to ecosystems are most often valued using contingent valuation estimates of existence value or direct ecosystem services (e.g., Nordhaus and Boyer, 2000; Anthoff and Tol, 2014). It is important, however, to note that damages to ecosystems may amplify other impacts. For instance, vegetation affects hydrology (e.g., Davie et al., 2013). As another example, about one-third of global agricultural production depends on animal pollination (Klein et al., 2007), so the loss of diverse animal pollinators as a result of climate-driven ecosystem stress could aggravate impacts of climate change on agriculture. Similarly, reductions in biodiversity can promote the spread of vector-borne diseases (LoGiudice et al., 2003), which is also influenced by climate (e.g., Altizer et al., 2013; Caminade et al., 2014). For the damages module in general, hindcasting and empirical calibration of models will be important tools for assessing the future representation of interactions and feedbacks.

Recognition and Consideration of Damages that Directly or Indirectly Affect Welfare

The individual sectoral impact functions available for inclusion in a damages module are estimated using a range of methods, as discussed above (see, especially, Table 5-3). Many are based on structural economic models of a sector or specific climate effect. A growing number of them derive empirical estimates by applying econometric methods to historical data, and some are processed through economic or integrated assessment models that may include various interactions among sectors or regions. There are differences in the information produced by these methods. In addition, there are important differences in the assumptions required to quantify different categories of climate change impacts. As a consequence, clarity regarding the underpinnings of the damage estimate requires transparency about the components of the estimate.

One important distinction is among damages that affect human consumption, those that affect capital stocks, and those that affect welfare in ways that are not mediated through markets. One output generated by many of the procedures underlying damage functions is an estimate of the net change in aggregate macroeconomic consumption of goods and services that are priced in markets. This measure of welfare is clear and flows directly into the discounting procedure and the SC-CO$_2$ estimate (see Figure 2-1, in Chapter 2). However, climate change does not always affect consumption directly, and may affect the level or productivity of capital stocks (physical, human, and environmental). Consumption effects are a downstream consequence of changes in input and output markets.

Impacts that harm capital stocks, the most well studied of which are the impacts of increased coastal flooding that affects durable infrastructure, will increase the demand for new investment. In the case of coastal flooding for example, this demand may divert investment from high-productivity activities to post-flood reconstruction and replacement of lost infrastructure. Using a CGE model, Bosello and colleagues (2007) found that the indirect costs of sea level rise, mediated by land loss or the capital market effects of protective investments, are comparable in scale to the direct effects. Using a CGE model, Houser and colleagues (2015) found that the long-term growth impacts of capital destruction caused by coastal storms on the United States as a whole were several times larger than the initial cost.

Effects on a particular type of capital stock will affect production input choices and markets, as well as output. For instance, Reilly and colleagues (2007) find that the macroeconomic effects of climate change are significantly smaller than the climate productivity shocks to land due to adaptation through markets, with changes in inputs, production, and international trade. Some of the effects of impacts on capital stocks may be captured in the functions estimating monetized consumption, but not necessarily all of them. There will be feedbacks and interactions among sectors that the available research does not yet capture. Therefore, to the extent possible, it will be important to take account of these capital stock effects as input to improved estimates of consumption and for possible consideration of feedbacks in sectoral interactions. In the longer term, incorporating these feedbacks to the socioeconomic module, discussed in Chapter 3, is of key importance.

Another potentially important welfare consequence of climate change is the loss of goods and services that are not traded in markets and so cannot be valued using market prices: examples include loss of cultural heritage, historical monuments, and favored landscapes; loss of charismatic and other species; violence; and forced migration. If kept in natural units, the distinction between estimates of these effects and those based on market prices will be transparent. However, some impacts may be treated as substitutable for consumption of market goods, and these effects may be converted into monetary terms using willingness to pay or other simulated market concepts. These nonmarket effects are an important consequence of climate change and need to be quantified in monetary terms to the extent possible. Because the assumptions underlying these estimates are fundamentally different from the assumptions that underlie procedures based on market prices, their role in any damage total needs to be made transparent.

Representing Adaptation to Climate Change and the Costs of Adaptation

Households, communities, and societies will each take action autonomously to reduce the welfare losses of a changing climate, and policy makers will also direct investment to adaptation. Understanding the effectiveness of such measures, and their cost, is part of understanding the SC-CO$_2$. For example, estimates of the costs of morbidity and mortality from extreme heat events will be overstated if they ignore greater use of air conditioning, but the overall damages must also include the cost of the greater use of air conditioning. In principle, the loss from the effect of a change in climate on some activity is the cost of adaptation measures plus the residual loss with the adaptation in place. In practice, such calculations can be analytically difficult.

The SC-IAM damage functions, and those in many other climate effects studies, represent climate damage as a function of global and regional mean temperature. However, climate change damages are often the effect of extreme events (e.g., a heat wave, storm, drought, or flood) involving other regional climate variables. Understanding of socioeconomic and ecological responses to these extreme events is limited, particularly at the relevant spatial scale, as is understanding of the relation of the change in these extremes to a projected change in global average temperature. This complexity not only creates difficulty for constructing estimates of climate damage, but also is a problem for the individuals and firms whose adaptation response is being modeled. Moreover, decision makers at all levels may have difficulty distinguishing between climate change and unforced weather variability, and their understanding may be further challenged by their own experiences and highly uncertain or conflicting projections from experts. As a result, they may take actions that are suboptimally early or late.

In spite of these complexities, it is important when constructing a new climate damage module to favor those damage estimates that take account of both adaptation (in order to avoid SC-CO$_2$ estimates that overstate potential future economic loss) and the costs of adaption. Calculation of these effects in some sectors is straightforward (e.g., changes in heating or cooling), yet they may be more complex as the adaptation response spills over into other sectors (e.g., the simultaneous effects of changes on heating or cooling on both health and energy consumption). Some structural economic models of climate impacts are well suited to consider the adaptation response. Based on historical experience, empirical models are likely to capture the adaptation that has occurred in the sector or location studied, but they will have a harder time extracting the adaptation response and its costs that are relevant to future, long-term changes that are not present in historical datasets. Advances in methods to consider adaptation responses may allow quantification of the costs of adaptation for a number of important sectors (e.g., agriculture, mortality) (e.g., Auffhammer and Aroonruengsawat, 2012; Butler and Huybers, 2013).

In contrast with process models, structural economic models can endogenously model future adaptation possibilities and their costs through changing markets (e.g., Reilly et al., 2007). Managed (e.g., policy-driven) and autonomous (e.g., market-driven) adaptation responses can be assessed in such a framework. Evaluation of the adequacy of damage estimates in capturing changes in vulnerability and success in adaptation would be a separate task for each damage function. Evaluation of overall performance would be limited to a rough assessment of the fraction of estimated damage for which explicit consideration of adaptation has been possible.

Representation of Nongradual Damages

The Earth system has the capacity to exhibit "abrupt," nonlinear shifts between states. Various terms are used to describe these discontinuous system dynamics: abrupt changes, critical thresholds, regime shifts, tipping points, surprises, discontinuities, and catastrophic events. This imprecise and inconsistent terminology complicates discussions of how these complex phenomena can be incorporated in damage estimates.

Potential "climatic tipping elements" that could exhibit such discontinuous dynamics include the Atlantic meridional overturning circulation (AMOC), monsoonal circulation patterns, sea ice, polar ice sheets, permafrost carbon, marine methane hydrates, and the Amazon rainforest (Alley et al., 2003; Lenton et al., 2008; National Research Council, 2013; Kopp et al., 2016b). Gradual changes in the physical climate may drive these tipping elements over a threshold, producing a new equilibrium state—such as one in which an ice sheet is dramatically smaller than today or the Amazon rainforest is a savannah, for example.

Outcomes with high consequences, even if they are unlikely, have the potential to dominate expected welfare changes (e.g., Weitzman, 2011); their omission could affect estimates of the SC-CO2. While difficult to estimate, the value of reducing the probability of high consequence events due to climate change could be quite large.

Many researchers point out that the SC-IAM damage functions fail to capture the risk of uncertain Earth system dynamics in an explicit or credible manner (Hitz and Smith, 2004; Warren et al., 2006; Kopp and Mignone, 2013; Deschenes, 2014; Howard, 2014; Li et al., 2014; Revesz et al., 2014; Sussman et al., 2014). Although the existence of these risks is supported by the geologic record (e.g., National Research Council, 2013) and in some cases by Earth system models (e.g., Drijfhout et al., 2015), the governing dynamics and thresholds are generally not well understood or quantified due to insufficient data and the limitations of process models. In addition, nongradual damages may arise from critical thresholds in socioeconomic systems as well as in natural systems. For example, by increasing the probability of civil conflict (Hsiang et al., 2013), gradual climate change could tip countries into a conflict-development trap, that is, is a self-reinforcing cycle in which civil conflict leads to slow or negative economic growth, and low economic development increases the risk of civil conflict (Collier et al., 2003).

The IWG needs to evaluate the state of knowledge and understanding of critical thresholds in climatic and climatically influenced socioeconomic tipping elements, as well as their likelihoods and consequences. It also needs to consider approaches for incorporating critical thresholds that can be appropriately quantified into the damage module. For example, Kopp and colleagues (2016b) propose an approach that includes using critical threshold scenarios in physical and empirical models to assess the potential impacts of crossing critical thresholds, together with structured expert elicitation to assess the probability of crossing those thresholds. A research program on critical thresholds, as well as on physical and economic modeling frameworks that incorporate them, would improve the capacity to integrate them into the SC-CO_2 estimation framework. Such a program is particularly needed because it is currently unknown whether there are critical thresholds whose crossing would lead to significant damages, including potential effects on economic growth that could also affect SC-CO_2 discounting (see Chapter 6).

CONCLUSION 5-1 An expansion of research on climate damage estimation is needed and would improve the reliability of estimates of the SC-CO_2.

- **In the near term, initial steps that could be undertaken include:**
  - **a comprehensive review of the literature on climate impacts and damage estimation, the evaluation of adaptation responses, and regional and sectoral interactions,**

as well as feedbacks among the damage, socioeconomic, and climate modules; and

-   a comparison of methods for estimating damages, including characterizations of their differences, synergies, uncertainties, and treatment of adaptation.

- In the medium to long term, several research priorities could yield particular benefits for SC-CO$_2$ estimation:

   -   physical, structural economic, and empirical estimation of climate impact relationships for regions and sectors not currently covered in the peer-reviewed literature;
   -   structural and empirical studies of the efficacy and costs of adaptation;
   -   calibration of damage functions using empirical and structural models operating at sufficiently high temporal and spatial resolution to capture relevant dynamics;
   -   the development of systematic frameworks for translating estimates of impacts into welfare costs; and
   -   empirical observation-based and structural modeling studies of interregional and intersectoral interactions of impacts, as well as of feedbacks among damages, socioeconomic factors, and emissions.

- In the long term, research priorities that could yield particular benefits for SC-CO$_2$ estimation would include omitted critical thresholds in natural and socioeconomic systems:

   -   development of simple Earth system model or full complexity Earth system model scenarios in which potential critical thresholds of tipping elements (e.g., Atlantic meridional overturning circulation, monsoonal circulation patterns, sea ice, polar ice sheets) are crossed, and the use of the physical changes in these scenarios to drive models that assess impacts and damages;
   -   empirical observation-based and structural modeling studies of the potential for climate change to drive the crossing of critical thresholds in socioeconomic systems and of their ensuing damages; and
   -   expert elicitation studies of the likelihood of different tipping element scenarios, in order to allow tipping elements and their critical thresholds to be represented probabilistically in the SC-CO$_2$ framework.

6

Discounting Module

Discounting is the process by which costs and benefits spread over current and future years can be compared in order to establish whether a particular choice leads to an overall net benefit. The discount rate refers to the reduction ("discount") in value each year as a future cost or benefit is adjusted for comparison with a current cost or benefit. This chapter first discusses the IWG's approach to discounting in the context of both broader government guidance and the academic literature on discounting, particularly regarding uncertainty about future economic growth. The second section looks broadly at approaches to discounting. The next two sections elaborate on likely correlations among climate damages, economic growth, and the appropriate discount rate and the idea that such correlations could be explicitly modeled in the $SC\text{-}CO_2$ estimation. The final section considers other discounting issues. Throughout the chapter, guidance is offered on future $SC\text{-}CO_2$ updates by providing examples of how they could be implemented and, more generally, how uncertainty about the discount rate could be handled.

IMPORTANCE OF DISCOUNT RATE ASSUMPTION FOR THE $SC\text{-}CO_2$ ESTIMATES

The discount rate plays an important role in estimating the $SC\text{-}CO_2$ because the impacts of today's CO_2 emissions persist and accumulate far into the future. The value today of avoiding those impacts depends heavily on how much society discounts those future impacts: small differences in the discount rate can have large impacts on the estimated SC-CO_2. This effect is highlighted in Table 1-1 (in Chapter 1), which shows the IWG estimated SC-CO_2 for discount rates of 2.5, 3.0, and 5.0 percent. The ratio of SC-CO_2 estimates based on 2.5 percent compared with 5 percent is a factor of up to five-fold (i.e., $10 versus $50 in 2010).

The underlying temporal trend of future discount factors and damages can be seen in Figure 6-1. This figure shows the committee's calculated patterns over time of discounting and of damages associated with three discount rates and one example of a damage scenario from an integrated assessment model (discussed in detail below). It is important to note that the scale is logarithmic. The top line shows the time profile of damages from a single ton of CO_2 emitted in 2015. The undiscounted damages rise from roughly 10 cents in 2015 to more than $100 in 2295. The discounted present value associated with $1 of future damages is indicated by the lower three lines for each of the three discount rates. For a 2.5 percent discount rate, this present value falls from $1 associated with $1 in damages in 2015 to less than one-tenth of 1 cent in 2295. For a 5 percent discount rate, $1 received in 2295 is valued at one-ten-thousandth of 1 cent today. This strikingly different result is due to the power of compounding discount rates over time.

To understand how the discount rates and damage estimate combine to form different SC-CO_2 estimates, Figure 6-2 shows the committee's computation of the present value of damages shown in Figure 6-1 using

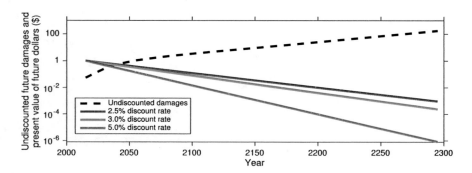

FIGURE 6-1 Undiscounted damages from 1 metric ton of CO_2 emissions in 2015 and present value of $1 received in the future using discount rates of 2.5, 3.0, and 5.0 percent.
NOTE: See text for discussion.

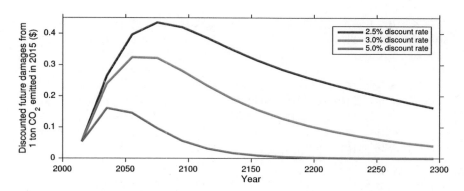

FIGURE 6-2 Pattern of discounted annual damages associated with a fixed pattern of undiscounted damages and three discount rates.
NOTE: See text for discussion.

each of the three discount rates.[1] The estimated SC-CO_2 from this damage example for each discount rate would equal the area under each curve. Two observations are immediately evident. First, for a given pattern of damages, the SC-CO_2 is much higher for low discount rates. Second, the modeling horizon needed to include most of the discounted damages varies with the discount rate. For this example time profile of damages, most discounted impacts are captured by 2150 when the discount rate is 5 percent. However, a significant amount of discounted damages may be missed even with a 300-year horizon when the discount rate is 2.5 percent. Another issue, though not apparent in Figure 6-2, is whether combining different discount rates with the same pattern of damages is always appropriate. As discussed in this chapter, the likely relationships between economic growth, discounting, and climate change damages is an important consideration.

The importance of the discount rate in benefit-cost analysis has not gone unnoticed. The U.S. Office of Management and Budget (OMB) has offered guidance on the use of particular discount rates dates for more than 40 years, and it has evolved over time (see, e.g., U.S. Office of Management and Budget, 1972). This guidance has been used in a wide range of regulatory analyses, ranging from food labeling to power plants.[2]

---

[1]The committee refers to the present value of a dollar received in year t as the discount factor for year t. The present values are computed by multiplying damages in year t by the discount factor for year t, using each alternative discount rate.

[2]See *Food labeling: trans fatty acids in nutrition labeling, Nutrient content claims, and health claims, Federal Register*, vol. 58, no. 133 (July 11, 2003) and U.S. Environmental Protection Agency (2015).

In order to present conclusions and recommendations about discounting applied to climate change damage estimates, the committee first discusses current OMB guidance and the scholarly literature on discounting. The IWG's approach and the justification for it, as well as how agencies have used the IWG values in regulatory impact analyses (RIAs) is then reviewed. In the Phase 1 report (National Academies of Science, Engineering, and Medicine, 2016), the desire for consistency in the use of discount rates in RIAs is discussed (see also Box 1-2 in Chapter 1). The committee returns to this issue below.

APPROACHES TO DISCOUNTING AND THEIR APPLICATIONS

The U.S. government approach to discounting, including both those of the OMB and the IWG, has largely rested on observed market rates. Both OMB guidance and the IWG (see Interagency Working Group on the Social Cost of Carbon, 2010) also discuss "prescriptive" approaches derived from a social welfare framework (detailed below). These approaches are briefly reviewed before turning to the specific issue of growth uncertainty and discounting over the long term.

OMB Guidance on Discounting

In RIAs of federal rules, the rate at which future benefits and costs are discounted can determine whether the net present value of a regulation or project is positive or negative. In accordance with OMB Circular A-4, for rules with both intra- and intergenerational effects, agencies traditionally use constant discount rates of 3.0 and 7.0 percent, as well as a possible lower rate to reflect important intergenerational costs and benefits. The rationale for the 7.0 percent rate is that it is an estimate of the average before-tax rate of return to private capital in the U.S. economy. The 3.0 percent rate is intended to reflect the rate at which society discounts future consumption, or "social rate of time preference," which is particularly relevant if a regulation is expected to affect private consumption directly (see U.S. Office of Management and Budget, 2003). A third, lower discount rate may be used as a sensitivity analysis if benefits or costs accrue to future generations over long time horizons.

OMB has provided more detailed rationales for these discount rates. In the return to capital approach, the discount rate is the rate of return on investment. This approach reflects the idea that, as long as the rate of return to capital is positive, society needs to invest less than $1 today to obtain $1 of benefits in the future. In the consumption approach, the discount rate reflects the rate at which consumers would be willing to trade

$1 of consumption today for a $1 of consumption in the future on the basis of the market tradeoffs that they face.

If all costs and benefits in an RIA are expressed in terms of their impacts on consumption, the appropriate discount rate is the consumption rate of interest.[3] If there were no inefficiencies or distortions in the economy, the average risk-adjusted rate of return on investment would equal the consumption rate of interest. There are, however, reasons why the two differ. For example, taxes on investment income imply that the return to private investment exceeds what is received after taxes by the consumer. It is also the case that the costs and benefits of a project are not always expressed in consumption equivalents. These factors are why OMB requires projects involving *intragenerational* benefits and costs to be evaluated using discount rates that reflect both approaches, as a sensitivity analysis.

The choice of a discount rate applied over longer time horizons raises questions of *intergenerational* equity. Whether the benefits of climate policies, which can last for centuries, outweigh the costs, many of which are borne in the nearer term, is especially sensitive to the rate at which future benefits are discounted. Although the influence of the discount rate on damages in the future is well understood, there is no consensus about what rate to use in the context of estimating the SC-CO$_2$ (Interagency Working Group on the Social Cost of Carbon, 2010).

Any rate used to estimate climate damages other than 3.0 or 7.0 percent presents complications in combining estimates of the SC-CO$_2$ with other benefit and cost estimates in an RIA. Specifically, using a constant discount rate for intergenerational benefits and costs that is lower than the rate used to evaluate intragenerational benefits and costs can lead to inconsistencies in decision making: consistency requires that the same discount rate must be applied to all benefits and costs that occur in the same year (Arrow et al., 2013). When uncertain outcomes are considered, the discount rate applied to costs and benefits in a given year may vary across uncertain outcomes but, for a particular outcome, they ought to be the same for all costs and benefits. The committee returns to this possibility below.

Descriptive and Prescriptive Approaches in the Literature

In the economics literature, two approaches are used to determine the appropriate discount rate in climate change analyses. The positive,

---

[3]"Interest rate" refers to measurable returns earned on various types of investment. As noted above, the discount rate refers to how one compares a dollar in the future with a dollar today—which may or may not equal various measurable returns.

"descriptive," approach rests on observed behavior in savings and investment decisions that individuals make in the real world. The normative,
"prescriptive," approach takes the perspective of a social planner who
prescribes weights to the welfare of future and current generations.

The descriptive approach focuses on setting the discount rate on the
basis of actual market rates of return. That is, the discount rate is inferred
from rates of return that reflect consumers' actual choices—for example,
savings versus consumption decisions or tradeoffs between more and
less risky investments (Interagency Working Group on the Social Cost
of Carbon, 2010). Three arguments are offered in favor of this approach:
(1) mitigation expenditures displace other forms of investment; (2) if the
return on mitigation investments is smaller than on other investments,
allocating resources to mitigation efforts may make current and future
generations worse off; and (3) it is preferable to base resource allocation
on the tradeoffs that society actually makes (Arrow et al., 1996).

Whether the descriptive approach calls for using the pretax return on
capital or the consumption rate of interest depends on whether benefits
and costs are measured in consumption equivalents. If they are, then the
theoretically correct discount rate is the rate at which consumers would
trade consumption today for consumption in the future. In many cases,
the benefits of avoiding climate change, such as health benefits, accrue
directly to consumers or affect the prices consumers pay for goods and
services. Even when climate damages do not directly affect consumers,
damage estimates from the SC-IAMs are reported in consumption-equivalent units.

In contrast, the prescriptive approach is based on a social welfare
function that reflects the weight that a policy maker attaches to the utility
of current and future generations. The discount rate under the prescriptive approach is the rate at which $1 received by a future generation must
be discounted to give the same marginal utility to the present generation
as it gives to the future generation. This discount rate, r, is given by the
following Ramsey formula,[4]

$$r = \delta + \eta \cdot g,$$

where δ is the discounting of the utility of future generations or "pure
time preference" rate; η is the change in the value of an additional dol-

---

[4]The approach was pioneered by Frank Ramsey (1928), with many extensions and elabora-
tions. An important limitation of this approach is that η conflates risk aversion and the intertemporal elasticity of substitution. While the latter is our main focus here, future research
could explore alternative formulations that relax this restriction, such as along the lines of
the Epstein-Zin preferences (Epstein and Zin, 1989, 1991; Ackerman et al., 2013).

lar as society grows wealthier (the absolute value of the "elasticity of marginal utility of consumption"); and g is the growth rate of per capita consumption.[5]

An implication of the Ramsey equation is that the discount rate is inherently linked to the growth rate of the economy. This interdependence suggests that the rate used to discount future climate damages needs to be consistent with assumptions about the rate of economic growth that underlie the emissions path in the socioeconomic module and the calculation of climate damages in the damages module.

Arguments for adopting the Ramsey-based welfare approach to discounting include the notion that the discount rate ought to be derived from ethical considerations reflecting society's views concerning consumption tradeoffs across generations. It is also true that there are few market interest rates that provide indicators of consumption tradeoffs over horizons longer than a few decades.[6]

In parameterizing the Ramsey formula, the *Fifth Assessment Report* (AR5) of the Intergovernmental Panel on Climate Change (IPCC) (Kolstad et al., 2014) and the IWG *Technical Support Document* (Interagency Working Group on the Social Cost of Carbon, 2010) provide a synthesis of the relevant literature, which suggests the following parameter values:

- Pure time preference rate, δ: Many papers in the climate change literature adopt values in the range of 0 to 3 percent per year (Interagency Working Group on the Social Cost of Carbon, 2010), although the largest value cited in the AR5 (Kolstad et al., 2014) is 2 percent, with the majority of values cited equaling zero or a number close to zero. One argument for a value of δ equal to 0 is that, holding consumption constant, all generations ought to be given equal weight in calculating social welfare.[7]
- Elasticity of marginal utility of consumption, η:[8] Also referred to as intergenerational inequality aversion, the value of η typically falls in the range of 1 to 4 (Kolstad et al., 2014).
- Growth rate of per capita consumption, g: A commonly used

---

[5]Note that while g is per capita consumption growth, the discount rate is applied to total (not per capita) benefits and costs because welfare depends on the total population.

[6]For example, the longest terms for U.S. Treasury bonds and most home mortgages is 30 years. Very few private markets provide evidence about longer-term rates: (see Giglio et al., 2015).

[7]Sometimes a small positive rate is used to account for the probability of human extinction due to causes unrelated to climate change (see, e.g., Stern, 2007).

[8]The elasticity of the marginal utility of consumption with respect to consumption is negative; hence, η represents the absolute value of the elasticity of marginal utility with respect to consumption.

estimate in the recent literature for g is 2 percent per year, based on global growth over the past few decades (see Appendix D for a discussion of global growth data and projections).

While g is determined by the performance of the economy and is observable (ex post), δ and η are never observable, but require an ethical judgment. Some studies make judgments directly regarding the magnitude of δ and η (e.g., $\delta = 0$). Other studies assume observed individual behavior can inform social preferences and proceed to estimate (or calibrate) either δ or η from empirical evidence. But even in the latter case, it is an ethical judgment to conclude that societal values are defined by individual behavior. Moreover, η can be associated with risk aversion, aversion to inequality across individuals in a given generation, and aversion to uneven consumption over time for an individual—as well as inequality aversion across generations. Furthermore, some studies take a descriptive approach and choose δ and η to calibrate the Ramsey equation to market rates. Estimates of η based on these different notions differ considerably (e.g., Atkinson et al., 2009). Thus, further judgement is required to choose among various estimates. The AR5 (Kolstad et al., 2014) summarizes a variety of such efforts spanning both academic research and government policy making. It identifies a range of implied discount rates from 1.4 to 6 percent: see Table 6-1.

Uncertainty about Future Discount Rates

Over long time horizons, the discount rate is uncertain. This is true under the descriptive approach because future market rates of interest are uncertain. It is also true under the prescriptive approach because future growth rates are inherently uncertain. In both approaches, discounting when rates are uncertain is more complex than simply using an expected or average discount rate.

Suppose under the descriptive approach that net benefits at time t, $Z(t)$, are discounted to the present at a constant exponential rate r, so that the present value of net benefits at time t equals $Z(t)\exp(-rt)$.[9] If the discount rate r is fixed over time but uncertain, then the expected value of net benefits is given by $E(\exp(-rt))Z(t)$.[10] The certainty-equivalent discount rate, R_t, used to discount $Z(t)$ to the present, is defined by

---

[9]This assumes that $Z(t)$ represents certain benefits. If benefits are uncertain we assume that they are uncorrelated with r and that $Z(t)$ represents certainty-equivalent benefits. The case of uncertain benefits is further discussed below.

[10]In this chapter, we use $E[.]$ to represent the expectation operator: that is, it represents the mean value of the random variable in brackets.

TABLE 6-1 Values and Implied Social Discount Rates in Selected Studies

Author	Rate of Pure Time Preference (in %)	Risk/Inequality Aversion	Anticipated Growth Rate (in %)	Implied Social Discount Rate (in %)
Cline (1992)	0.0	1.5	1.0	1.5
IPCC (1996)	0.0	1.5-2.0	1.6-8	2.4-16
Arrow (1999)	0.0	2	2.0	4.0
UK: Green Book (HM Treasury, 2003)	1.5	1	2.0	3.5[a]
U.S. Office of Management and Budget (2003)[b]				3-7
France: Rapport Lebègue (2005)	0.0	2	2.0	4.0[a]
Stern (2007)	0.1	1	1.3	1.40
Arrow (2007)		2-3		
Dasgupta (2007)	0.1	2-4		
Weitzman (2007)	2.0	2	2.0	6.0
Nordhaus (2008)	1.0	2	2.0	5.0

NOTES: The table shows the calibration of the discount rate based on the Ramsey rule; see text for discussion.
[a]Decreasing with the time horizon.
[b]OMB uses a descriptive approach.
SOURCE: Adapted from Intergovernmental Panel on Climate Change (2014c, Table 3-2).

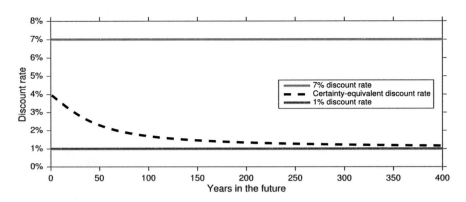

FIGURE 6-3 Certainty-equivalent discount rate for different horizons.
SOURCE: Committee generated.

$$\exp(-R_t t) = E(\exp(-rt)). \tag{1}$$

That is, R_t represents the certain discount rate that is equivalent to an uncertain discount rate in terms of the discount factor over a particular horizon (which is what matters for cost-benefit analysis).[11] As shown in Figure 6-3, if $r = 1$ percent or 7 percent, each with probability 0.5, the certainty-equivalent discount rate decreases from 3.96 percent for a 1-year horizon to 1.69 percent for a 100-year horizon, to 1.17 percent for a 400-year horizon. The convexity of the discount factor guarantees that the certainty-equivalent discount rate is always less than $E(r)$ and that it declines over time.[12]

In the more general case in which the discount rate is uncertain and varies over time, the expected discount factor is given by $E[\exp(-\Sigma_{\tau=1...t} r_\tau)]$. In this case, the shape of the R_t path depends on the distribution of the per-period discount rates, $\{r_\tau\}$. If $\{r_\tau\}$ are independently and identically distributed, the certainty-equivalent discount rate is constant. There must be persistence in uncertainty about the discount rate for the certainty-

---

[11]The literature sometimes refers to a certainty-equivalent "forward rate" equal to the appropriate rate to discount certain values between two adjacent future periods: that is, $E(\exp(-r(t + 1)))/E(\exp(-rt)) = \exp(-R_{t+1}(t + 1))/\exp(-R_t t)$. This forward rate can thus be written in terms of our certainty-equivalent rate as $(t + 1)R_{t+1} - tR_t$.

[12]This result is guaranteed by Jensen's inequality, which states that the expectation of a convex function is always larger than the function of the expectation. Formally, $E(\exp(-rt)) > \exp(-E(r)t)$.

equivalent rate to decline. If, for example, shocks to the discount rate are correlated over time,

$$r_t = \pi + e_t \text{ and } e_t = ae_{t-1} + u_t, |a| \leq 1, \qquad (2)$$

where π and a are fixed parameters and $\{u_t\}$ are independently and identically distributed random variables, then the certainty-equivalent discount rate will decline over time if $a > 0$ (Newell and Pizer, 2003).

In the Ramsey formula, uncertainty in the rate of growth in consumption can likewise lead to a declining certainty-equivalent discount rate. Arrow and colleagues (2014) note that the standard Ramsey formula for the consumption rate of discount can be extended to handle uncertainty about the rate of growth in consumption by subtracting a third "precautionary" term (Mankiw, 1981; Gollier, 2002). If growth is subject to independently and identically distributed shocks, this term will reduce the discount rate, but not cause it to decline.[13]

If random shocks to growth are positively correlated over time, however, the precautionary term in the Ramsey formula may become sizable in absolute value for long horizons, leading to a declining term structure of discount rates (see Gollier, 2012, for an extended survey).[14] Uncertainty about the mean and variance of the rate of growth in consumption can also lead to a declining risk-free discount rate, rather than a constant exponential rate as used by the IWG (Weitzman, 2004, 2007; Gollier, 2008).

The IWG's Approach

In estimating the SC-CO$_2$, the IWG relies on guidance from OMB Circular A-4 and the economics literature to defend the use of a consumption rate of interest as the appropriate rate for discounting the net benefits of a marginal change in carbon emissions. The estimates that result from the SC-IAMs[15] are measured in consumption-equivalent units: thus, a discount rate that reflects how individuals trade off current and future consumption is defensible in this setting.

The specific consumption rate of interest used to discount future

---

[13]For independently and identically normally distributed shocks with variance σ^2 and a mean growth rate of $E(g)$, the certainty-equivalent discount rate r will be $r = \delta + \eta E(g) - 0.5\eta^2\sigma^2$.

[14]That is, the appropriate rate at which to discount a quantity at some future time t to the present declines as t grows.

[15]These are the three integrated assessment models widely used to produce estimates of the SC-CO$_2$: Dynamic Integrated Climate-Economy model (DICE), Framework for Uncertainty, Negotiation and Distribution model (FUND), and Policy Analysis of the Greenhouse Effect model (PAGE); see Chapter 1.

climate damages depends on the correlation between damages and consumption. The exact value of future climate damages is inherently uncertain. So long as these damages have little correlation with the growth of consumption, it is appropriate to discount expected damages at a risk-free rate. That is, one would use a discount rate associated with either the expected growth rate under a prescriptive, Ramsey approach or a relatively low-risk bond (e.g., U.S. government bonds) under a descriptive approach. Alternatively, if damages are positively correlated with consumption, the discount rate would be larger, and if they are negatively correlated with consumption, the discount rate would be smaller than the risk-free discount rate.

Existing OMB guidance on discounting does not fully address the issue of discounting over long horizons or the effect of uncertainty on discount rates, both of which directly influence the SC-CO$_2$ estimates. The IWG made modifications to adapt the OMB guidelines to reflect these points. Specifically, the IWG chose three constant, exponential annual discount rates (2.5, 3.0, and 5.0%) and presented results conditional on each of these discount rates.

The central value of a 3.0 percent rate, consistent with the consumption rate of interest in OMB Circular A-4 guidance, is meant to reflect the post-tax, risk-free interest rate. The 5.0 percent rate is included to represent the possibility that climate damages are positively correlated with consumption growth. Uncertain investments with a high payoff in better times, and low payoff in worse times, are less valuable and require a higher rate of return than investments without such correlation. This would be the case if most of the impacts of climate change increase with the size of affected market sectors, such as real estate, agriculture and energy, or affect nonmarket sectors such as ecosystem quality or health, for which willingness to pay typically increases with consumption levels, thus leading to a positive correlation between the net benefits from climate policies and market returns (Interagency Working Group on the Social Cost of Carbon, 2010). Lastly, the 2.5 percent rate is intended to reflect uncertainty in the discount rate itself, as discussed in the previous section, as well as possible negative correlation between climate damages and consumption (i.e., the opposite of the rationale for 5.0%). The rate is based on the average certainty-equivalent rate of the random walk and mean-reverting models posited by Newell and Pizer (2003). This approach utilizes observed interest rates on Treasury notes to measure the risk-free consumption rate of interest (Interagency Working Group on the Social Cost of Carbon, 2010) and assumes no correlation between damages and the discount rate. Notably, the majority of climate change impacts studies cited in the AR5 use an implied social discount rate of no more than 5 percent (Kolstad et al., 2014).

In the executive summary of the *Technical Support Document*, the IWG presents results conditional on each of the three assumed discount rates for different years of emissions (see Figure 1-1 in Chapter 1). The SC-CO$_2$ per metric-ton of CO$_2$ emitted in 2020 is \$12 using a 5.0 percent discount rate; \$42 using a 3.0 percent discount rate; and \$62 using a 2.5 percent discount rate (all in 2007 dollars). This comparison highlights the importance of the choice of discount rate on SC-CO$_2$ estimates: the SC-CO$_2$ estimate for the central discount rate (3.0 percent) is more than three times the magnitude of the estimate using largest discount rate (5.0 percent).

CONCLUSION 6-1 In the current approach of the Interagency Working Group, uncertainty about future discount rates motivates the use of both a lower 2.5 percent rate and higher 5.0 percent rate, relative to the central 3.0 percent rate. However, this approach does not incorporate an explicit connection between discounting and consumption growth that arises under a more structural (e.g., Ramsey-like) approach to discounting. Such an explicit analytic connection is especially important when considering uncertain climate damages that are positively or negatively associated with the level of consumption. The Ramsey formula provides a feasible and conceptually sound framework for modeling the relationship between economic growth and discounting uncertainty.

Discounting Climate Benefits in RIAs

In RIAs that use SC-CO$_2$ estimates to quantify climate benefits, there are two typical approaches to discounting: a "snapshot" year and a cumulative net present value.[16] A snapshot year approach calculates the change in CO$_2$ emissions occurring in a given year (e.g., 2030) and discounts the reduction in future damages that accrue from those marginal emissions changes back to 2030. In practice, this means multiplying these emission changes by an SC-CO$_2$ estimate for that year. Other costs and benefits are then computed for effects of other policy-induced changes in 2030, including benefits from non-CO$_2$ emission reductions in 2030 that may accrue in future years. These benefits are combined with the estimated change in CO$_2$ mitigation benefits. The result is a "snapshot" of net ben-

---

[16]For an example of the snapshot year approach, see the RIA for the Clean Power Plan Final Rule from the U.S. Environmental Protection Agency (EPA): https://www.epa.gov/sites/production/files/2015-08/documents/cpp-final-rule-ria.pdf [January 2017], pp. ES-19 through ES-23. For an example of the net present value approach, see the RIA for EPA's CAFE Standards Final Rule: https://www.regulations.gov/document?D=EPA-HQ-OAR-2009-0472-11578 [January 2017], pp. 7-127 through 7-134.

efits associated with all (CO_2 and non-CO_2) emission changes in 2030. With this approach, a series of snapshot years are often chosen, with CO_2 mitigation benefits combined with other cost and benefit estimates for policy changes in each of those snapshot years. If this approach is used, costs and benefits in each snapshot year are not typically discounted back to present day and combined. In contrast, a net present value approach effectively does the same thing, but then computes a net present value of net benefits across snapshot years.

In most RIAs, different discount rates are used to compute the costs and benefits of different emission changes in each snapshot year. The discount rates applied to CO_2 benefits from emission changes in a snapshot year are 2.5, 3.0, and 5.0 percent (plus the 95th percentile for the 3.0 percent rate), following guidance from the IWG *Technical Support Document* (Interagency Working Group on the Social Cost of Carbon, 2010). Meanwhile, the discount rates applied to benefits from other emission changes in a snapshot year are 3.0 and 7.0 percent, the standard rates from OMB Circular A-4. Estimates are calculated for each of these benefit-discount rate combinations in each snapshot year. Not all of these estimates, however, are presented in summary material for the RIAs. In the Clean Power Plan Final Rule, for example, only the CO_2 benefits for a 3.0 percent discount rate are presented in the executive summary.

Similarly, when discounting climate and nonclimate benefits back to the present day under the cumulative net present value approach, discount rates remain consistent within benefits categories. That is, discounted damages for some future snapshot year are discounted back to the present using the same rates used to discount to the snapshot year. The choice of discount rates used is determined, essentially, by whether one is discounting climate or other benefits.

Both approaches illustrate the challenge of combining cost and benefit estimates when only some categories of cost and benefits have an intergenerational component. Absent an intergenerational component, OMB instructions to discount using 3.0 and 7.0 percent can be viewed as striking a balance between simplicity and analytical rigor. This intragenerational context represents the vast majority of applications. In an intergenerational context, however, OMB itself recognizes that the simple approach is insufficient and that additional ethical considerations arise. Confronting these issues and concerns in the SC-CO_2 context leads to the use of generally lower discount rates, but it leaves unresolved how they might be combined with intragenerational costs and benefits.

LINKING UNCERTAINTY IN DISCOUNT RATES
AND UNCERTAINTY IN ECONOMIC GROWTH

As noted in the above discussion, persistent uncertainty about future discount rates mathematically leads to a declining certainty-equivalent rate, which is the rate at which a certain benefit at time t would be discounted to the present. A considerable literature has grown up around this issue and demonstrated that such declining rates arise regardless of whether discounting uses a descriptive or prescriptive approach (Arrow et al., 2014; Cropper et al., 2014).

In the IWG approach, 3.0 percent has been used as a central value, motivated by the average risk-free rate measured over a very long period. An alternative low value of a 2.5 percent rate was largely motivated by this uncertainty and the declining rate argument. The IWG is not alone in this consideration. Both the United Kingdom and France have adopted declining discount rates for cost-benefit analysis based on these arguments.

As one confronts the reality that future discount rates are uncertain, an important complication is that the discussion of declining rates applies in its simplest form to a certain flow of costs and benefits. Alternatively, the costs and benefits being discounted may be uncertain but uncorrelated with any uncertainty about the discount rate. That is, suppose one is attempting to compute

$$E[\exp(-rt)X_t],$$

where r is an uncertain discount rate, and X_t is an uncertain climate change impact. It is correct to rewrite that as

$$E[\exp(-rt)]E[X_t]$$

if r and X_t are uncorrelated. But if they are correlated, a covariance term arises: it will be a negative effect in the case of positive correlation, lowering the expected net present value of damages, and positive in the case of a negative correlation, thus raising it. For a variety of reasons discussed below, uncertain future climate change impacts may well be correlated with uncertain future discount rates. Before discussing this point, we further explore why the IWG used a related line of thinking to argue for use of a 5.0 percent rate (and to provide an additional motivation for a 2.5 percent rate).

Correlation between Impacts and Discounting

One important reason for potential correlation between damages and discounting is that damages directly related to economic activity are tied to the overall size of the future economy, while the value of impacts on human health and mortality are likely tied to future per capita consumption levels. Both of these relationships exist in the current SC-IAM damage formulations (see Chapter 5). Even if future climate damages were relatively certain in terms of the fraction of pre-damage consumption levels, they would still be strongly correlated with uncertain economic growth (possible countervailing effects are discussed below).

Under a Ramsey approach to discounting, higher consumption per capita also implies greater discounting. Under a particular consumption growth scenario,

$$r_t(g_t) = \delta + \eta \cdot g_t \, ,$$

where r_t is the discount rate over t periods, that is, the rate used to discount net benefits in period t to the present period 0. The formula highlights that this discount rate is a function of g_t, the growth rate in consumption over the same t periods. As above, δ is the pure time preference rate and η measures how fast the marginal utility of consumption declines as consumption grows.

In perhaps the earliest integrated assessment under uncertainty, Nordhaus (1994b) explores alternative paths of economic growth rates. In this Ramsey-style model, the analysis implies both alternative magnitudes of climate impacts and alternative discount rates. In more recent work looking at the SC-CO_2 estimates, Nordhaus (2011) found little impact of growth uncertainty (or other uncertainty) on the SC-CO_2. He argues that low growth/low discounting scenarios are also low temperature/low damage outcomes. Even more recently, Nordhaus (2014) reframes this result as emphasizing the importance of $r - g$ (what one might call "growth-adjusted" discounting) for the SC-CO_2 estimates when marginal damages scale directly with economic activity and growth. For η near 1 and climate damages roughly proportional to total consumption, $(r - g)$ is relatively constant over various consumption growth rates, and so is the SC-CO_2.[17] As an alternative, one could imagine increased climate resilience at higher incomes leading to lower, possibly negative correlation between economic growth and damages. Without drawing conclusions about the specific relationship between damages and economic growth,

---

[17]For $\eta = 1$ and damages exactly proportional to total consumption, as in DICE, the dependency of discounted damages on the size of the economy is removed entirely, and discounting is determined entirely by the pure time preference rate.

this argument makes clear the *potential* correlation between discounting and damages can make a difference in SC-CO$_2$ estimates.

There is a second important reason to consider correlation of climate change impacts and discount rates. The potential for catastrophic impacts raises the possibility that some uncertain outcomes may involve much lower rates of economic growth and higher incremental damages *because* of climate change (Sandsmark and Vennemo, 2007; Kopp et al., 2012; Murphy and Topel, 2013). The implication would be a higher expected present value of damage than if the correlation is ignored. However, making this argument operational requires an integrated assessment model with a well-specified model of catastrophic damages.

It is less clear what relationship ought to exist between economic growth and discounting under a descriptive approach. One can write out the Ramsey relationship without interpreting the parameters in terms of welfare. That is, it is possible to imagine interest rates varying with the rate of per capita consumption growth without deriving the Ramsey equation from an optimal growth model. The historical evidence on the correlation between consumption growth and market interest rates is, however, difficult to interpret. Hall (1988) was one of the first to examine this question, and he found little correlation over time between short-term consumption growth and interest rates in the aggregate data. Examining term structures, however, Harvey (1988) noted that future growth is higher when longer-term rates exceed short-term rates, which suggests that long-term rates (or their difference from short-term rates) are correlated with future growth. Of course, there is only limited evidence on the term structure over multicentury time horizons (Giglio et al., 2015).

Gollier (2014) provides an alternative framework for considering the same set of issues through a standard consumption-based capital asset pricing model. In this framework, the appropriate rate for discounting future climate change impacts is

$$r_t = r_{ft} + \beta\pi_t ,$$

where r_t is the discount rate used to discount period t to the present, r_{ft} is the risk-free rate over this period, π_t is a measure of uncertainty about future average consumption growth over this period, and β is a measure of how future climate impacts vary with consumption.[18] As above, persistent uncertainty about consumption growth leads to a declining risk-free term structure, reflected in a declining value of r_{ft} over longer horizons. That same persistent uncertainty about consumption growth

---

[18]Specifically, the Gollier (2014) model assumes climate impact at time t is proportional to c_t^B.

will also lead to a rising risk premium, reflected in a rising value of π_t over longer horizons. The correlation between consumption growth and climate impacts reflected in β can lead to a rising or falling term structure, depending on the sign of the correlation. Specifically, Gollier (2014) shows that if $\beta > \eta/2$, the net effect is a rising term structure. As Gollier and Hammitt (2014) note, whether one effect or the other is dominant is "exploratory and controversial."

> RECOMMENDATION 6-1 The Interagency Working Group should develop a discounting module that explicitly recognizes the uncertainty surrounding discount rates over long time horizons, its connection to uncertainty in economic growth, and, in turn, to climate damages. This uncertainty should be modeled using a Ramsey-like formula, $r = \delta + \eta \cdot g$, where the uncertain discount rate r is defined by parameters δ and η and uncertain per capita economic growth g. When applied to a set of projected damage estimates that vary in their assumptions about per capita economic growth, each projection should use a path of discount rates based on its particular path of per capita economic growth. These discounted damage estimates can then be used to calculate an average SC-CO$_2$ and an uncertainty distribution for the SC-CO$_2$, conditional on the assumed parameters.

Practical Assessments of the SC-CO$_2$ with Uncertain Outcomes, Economic Growth, and Correlation

Representation of the uncertainties and their interrelationships through Monte Carlo simulations allows explicit exploration of the implications of discount rate uncertainty for the discounting of future climate change impacts. Choosing particular values for δ and η leads to a particular value for the risk-free discount rate over a given time period conditional on economic growth (g_t). Simulating uncertain pathways for economic growth can thus generate a term structure for the risk-free rate $r_{ft}(g_t)$. That is, it can produce the rate appropriate for a stream of certain climate impacts or for climate impacts that are uncorrelated with economic growth.

However, it is possible to do more: specifically, it is possible to simulate climate change outcomes for each g_t pathway. For each Monte Carlo simulation, the discounted SC-CO$_2$ contribution from each period can then be computed using the value of $r_{ft}(g_t)$ that corresponds to that g_t pathway. Then, the SC-CO$_2$ itself can be computed by averaging discounted SC-CO$_2$ contributions across simulations and adding over periods. It is also possible to infer the discount rate term structure for climate change

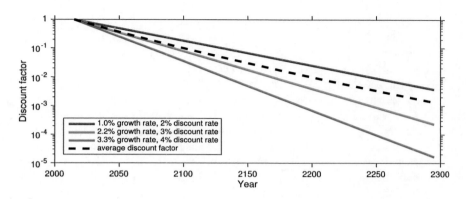

FIGURE 6-4 Examples of discount factors from 2015 to 2295.
NOTE: See text for discussion.

impacts by comparing the average discounted SC-CO$_2$ contribution from each period to the average undiscounted contribution.[19]

As an example, consider the following stylized model using the three, equal-probability economic growth scenarios described in Chapter 3, with constant per capita growth rates of 1.0 percent, 2.2 percent, and 3.3 percent. Also assume, for example, that δ = 1.1 percent and η = 0.88. Based on the Ramsey formula, the appropriate discount rates for the 1.0 percent, 2.2 percent, and 3.3 percent growth scenarios are, respectively, 2.0 percent, 3.0 percent, and 4.0 percent. With these assumptions, the committee calculated discount factors for each growth scenario, which are shown in Figure 6-4, along with the average discount factor for each future year assuming equal weights on the scenarios.

Importantly, representing uncertainty about the appropriate discount rate with multiple scenarios, each having a different constant discount rate, implies a declining discount rate. That is, in calculations of the present value of a future certain value (e.g., of damages) or an uncertain value that is uncorrelated with economic growth, the effective discount

---

[19]If agencies continue to use the net present value (NPV) approach for RIAs, it requires discounting the SC-CO$_2$ associated with emissions in each future year back to the current year. This could be accomplished in several ways. The IWG could present a table of the SC-CO$_2$ for each future year to be used for calculations using the snapshot year approach, as well as a table of the SC-CO$_2$ in current year dollars for calculations using the net present value approach. Alternatively, the IWG could present a table of appropriate discount factors derived from the discount rate term structure for climate impacts noted above. Yet another alternative is that the IWG could suggest using the near-term certainty-equivalent rate associated with each SC-CO$_2$.

rate declines over longer horizons. This outcome can be seen by using the average discount factor from Figure 6-4 to compute the corresponding certainty-equivalent discount rate—the rate that would be used to discount damages in each period back to 2015.[20] This is shown in Table 6-2 for the illustrative example in Figure 6-4.

The decline in the certainty-equivalent rate from 3.0 percent in 2015 to 2.4 percent in 2295 is a direct implication of allowing the rate of per capita consumption growth to be uncertain. Rather than using uncertain future discount rates to motivate a lower, fixed discount rate as the IWG did in its rationale for a 2.5 percent rate, allowing the rate of growth in consumption to be uncertain explicitly models that behavior. This approach implies a declining effective discount rate over long horizons for known future values or values uncorrelated with economic growth.

Consideration of Correlation of Discounting, Economic Growth, and Climate Damages

To incorporate climate change damages in the committee's example, imagine that other assumptions about population, emissions, climate change, and impacts yield the pattern of incremental damages over time from 1 metric ton of CO_2 emitted in 2015 for each of these three growth scenarios shown in Figure 6-4. Note that in this particular example, damages are positively related to economic growth: see Figure 6-5a.[21]

As noted throughout this report, there are many sources of variation in damages distinct from variation in economic growth. There may be many more scenarios than the three in Figure 6-5a, but each would have an associated path of economic growth rates. To illustrate how this discount rate schedule could be implemented in practice, each of these three projected damage estimates would be discounted on the basis of a discount rate defined by the assumed growth rate path in that projection. For the committee's example, one would multiply each projection of the damages in Figure 6-5a by the corresponding projection of discounting Figure 6-4: the result is shown in Figure 6-5b.[22]

To construct a valid SC-CO_2 estimate, the values for each scenario are then summed: in our purely illustrative example, this would yield

---

[20]To illustrate, in Figure 6-4, the average discount factor for 2035 is 0.56. The certainty-equivalent discount rate r_{f20} is the solution to the equation: $0.56 = \exp(-20 * r_{f20})$; in this case, $r_{f20} = 0.029$.

[21]The committee used a version of the DICE model to generate these damages: the key feature is that damages scale almost exactly with economic activity.

[22]To illustrate, for each year, the damage based on a 1.0 percent growth rate in Figure 6-5a (the blue curve) is multiplied by the corresponding discount factor in Figure 6-4, given by the blue line.

TABLE 6-2 Expected Discount Factor Based on Example Scenarios and Corresponding Certainty-Equivalent Discount Rates

Discount Factors and Discount Rates	2015	2035	2055	2075	2095	2115	2135	2155	2175	2195	2215	2235	2255	2275	2295
Average Discount Factor	1.00	0.56	0.32	0.19	0.11	0.071	0.044	0.028	0.018	0.011	0.0074	0.0048	0.0032	0.0021	0.0014
Certainty-Equivalent Rate (in %)	3.0	2.9	2.9	2.8	2.8	2.7	2.6	2.6	2.6	2.5	2.5	2.5	2.4	2.4	2.4

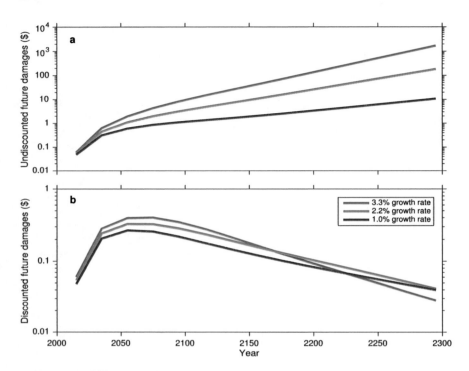

FIGURE 6-5 Example of undiscounted and discounted future damages from 1 incremental ton of CO_2 in 2015.
NOTES: The three growth scenarios are shown in undiscounted dollars. The population projection is fixed for simplicity.
SOURCE: Committee generated.

SC-CO_2 estimates in the range of \$38, \$47, and \$52 per ton, respectively, for the 1.0, 2.2, and 3.3 percent economic growth scenarios. The average of these values (and others) would yield a central SC-CO_2 estimate and the distribution used to describe a reasonable range of uncertain values. It is interesting to note that the estimates in each of the three scenarios are similar in this example. The lower discount factors (associated with higher discount rates) for high-growth scenarios largely offset the higher damages associated with those same high-growth scenarios.

Conversely, consider a case in which income is protective against climate damages. As an extreme example, suppose in the above case that, once global GDP reaches five times its current level, the world is sufficiently technologically advanced to eliminate all future climate damages. In this case, the SC-CO_2 estimates for the three economic growth scenarios would be $30, $19, and $15 per ton, respectively, for the 1.0, 2.2, and 3.3 percent economic growth scenarios; the lower discount factors (associated with higher discount rates) for high-growth scenarios would exaggerate the difference between the scenarios.

A more complicated scenario would occur if climate change damages were sufficiently large, with some probability, to alter the path of economic growth to an appreciable extent. As noted earlier, this would require a model of catastrophic damages that feed back to economic growth.

PARAMETERIZATION OF A RAMSEY-LIKE APPROACH FOR SC-CO$_2$ DISCOUNTING

The preceding discussion describes how one might simulate the combined effects of uncertainty on discounting and damages in a consistent way, but it does not address the question of how the parameters of a Ramsey-like approach might be chosen. Motivated by the Ramsey model itself, one could look to empirical assessments of pure time preference and utility curvature. That is, one could make an ethical judgment that social preferences would reflect individual preferences revealed through individual behavior. For example, the Intergovernmental Panel on Climate Change (2014c) report suggested a range of possible values of 0-2 percent for pure time preference (δ) and of 1-4 for risk/inequality aversion (η) as noted in Table 6-1 (above), many of which were driven by such an approach.

It is worth noting that the simulations described in the preceding section with δ and η chosen as normative welfare parameters may not lead to rates that are comparable to observed discount rates. Alternatively, one could choose those parameters to match empirical features of observed interest rates and the long-term relationship between interest rates and economic growth (as in Nordhaus, 2014). For example, given an initial economic growth rate, it is possible to pick combinations of δ and η that match, for initial years of the SC-CO$_2$ calculation, the central value of 3 percent used by the IWG. Among the locus of such values, there will be a range of resulting term structures for the risk-free rate describing how the discount rate changes after the initial years. That is, all would start at 3 percent, but would decline in the future based on the uncertainty surrounding future economic growth and the choice of parameters. For example, picking $\delta = 3$ percent and $\eta = 0$ would yield a flat term structure.

One can use the above simulation framework (with equally probable growth rates of 1.0, 2.2, and 3.3 percent) to examine the implications of alternate values of δ and η for the term structure. The above simulation assumed $\delta = 1$ percent and $\eta = 0.88$, yielding a 2015 certainty-equivalent rate of 3 percent and a 2295 rate of 2.4 percent (see bottom row of Table 6-2). Note that $\delta = 3$ percent and $\eta = 0$ yields a flat term structure with both the 2015 and 2295 rates being 3 percent. A few other experiments the committee calculated are shown in Table 6-3.

In the example in the preceding section, a key choice is what near-term rate to match. A potential guidepost is the regulatory framework in which the SC-CO$_2$ will be used. OMB provides considerable guidance concerning discount rates and their use in regulatory analysis, particularly with respect to consumption-based impacts (e.g., the 3% rate).

In addition to choosing δ and η based on various considerations, there is a final consideration of how to present results. If one views δ and η as uncertain parameters in the SC-CO$_2$ calculation and it is plausible to consider their joint distribution, one could include them with other uncertain parameters. This would lead, ultimately, to a single mean estimate of the SC-CO$_2$, along with a distribution that might be summarized on the basis of a particular prediction interval. Alternatively, if one views δ and η as ethical or policy parameters, the treatment of uncertainty about these parameters needs to be distinct from the treatment of uncertainty about the growth rate. Moreover, current OMB guidance suggests specific sensitivity analysis with respect to the discount rate because of its potentially dramatic effect on policy evaluation. Therefore, one could present a set of SC-CO$_2$ estimates based on different specified values of δ and η. That is, conditional on several different choices of δ and η, a distribution of SC-CO$_2$ values (including an average value) could be presented to reflect other sources of uncertainty in the computation of the SC-CO$_2$.

> **RECOMMENDATION 6-2 The Interagency Working Group should choose parameters for the Ramsey formula that are consistent with theory and evidence and that produce certainty-equivalent discount rates consistent, over the next several decades, with consumption rates of interest. The IWG should use three sets of Ramsey parameters, generating a low, central, and high certainty-equivalent near-term discount rate, and three means and ranges of SC-CO$_2$ estimates.**

Given an average per capita GDP growth rate, E[g], Table 6-3 shows that a variety of δ and η values can target the implied near-term discount rate given by $r = \delta + \eta$ E[g]. Moreover, the certainty-equivalent rate remains relatively constant over several decades. While the long-term certainty-equivalent rates will vary over ranges for δ and η that are consistent with theory and evidence, the SC-CO$_2$ itself is likely to be less sensitive to alternative choices of δ and η if they are chosen to target

TABLE 6-3 Combinations of Pure Time Preference (δ) and Risk/Inequality Aversion (η) Consistent with a Discount Rate of 3.0 Percent in 2015

δ (%)	η	2015 Rate (%)	2055 Rate (%)	2295 Rate (%)
0.1	1.33	3.0	2.7	1.9
1.1	0.88	3.0	2.9	2.4
2.1	0.42	3.0	3.0	2.8
3.0	0.00	3.0	3.0	3.0

the same near-term discount rate.[23] When choosing values of δ and η to match a particular near-term rate, the IWG could test the sensitivity of the SC-CO_2 to alternative values and be especially wary of values that lead to nonconvergent sequences of discounted damages (i.e., damages that grow faster than the discount rate).

One way to construct low, central, and high estimates would be to consider empirical evidence on the possible range of consumption-based, near-term market rates (e.g., government bonds). Alternatively, one could consider other judgments about appropriate high and low values around a central, market-based estimate. In any case, however, note that our recommendation for three rates in no way endorses the targeting of a near-term 7 percent discount rate as the high rate or the targeting of a near-term 3 percent discount rate as the lowest sensible low rate.

First, a portion of the argument for a 7 percent discount rate rests on uncertainty and correlation with market returns. Our recommended approach accounts for these factors directly. Second, another portion of the argument for a 7 percent discount rate rests on the tax wedge between returns to investment and the net-of-tax return received by consumers. As highlighted by Bradford (1975) and Lind and colleagues (1982), the return to investment is the correct discount rate only under very restrictive assumptions. Finally, the notion put forward in our recommendation is that of a sensitivity analysis around a central value. An implication is that, if the central parameterization for discounting is associated with a near-term 3.0 percent rate, as in the current IWG approach, then the low and high values would be on either side of 3.0 percent.

RECOMMENDATION 6-3 The Interagency Working Group should be explicit about how the SC-CO_2 estimates should be combined in regulatory impact analyses with other cost and benefit estimates that may use different discount rates.

---

[23]This is not to say that there is no sensitivity and, particularly when considering the risk of catastrophic damages, the choice of η and δ may be more difficult.

The committee sees at least two options for how SC-CO$_2$ estimates based on Recommendation 6-2 (above) could be combined in RIAs with other cost and benefits estimates that may use different discount rates. First, the IWG could present high and low parameterization results as a sensitivity analysis meant to illustrate the effect on the SC-CO$_2$ and instruct agencies to focus only on the central parameterization in RIAs. Second, the IWG could suggest all three discount rate parameterization results be used when appropriate in RIAs. The central value could be used in situations in which only one discount rate scenario is presented for other RIA benefits and costs. When estimates of other costs and benefits are presented using multiple discount rates, there would need to be guidance on how to pair estimates that are based on a particular discount rate with the SC-CO$_2$ parameterization. Here, one could imagine several options: (1) combining other cost and benefit estimates with the SC-CO$_2$ estimate whose near-term discount rate most closely matches that particular discount rate; (2) combining other costs and benefits based on a high discount rate with the SC-CO$_2$ estimate based on its highest discount rate, and analogously combining the low discount rate estimates; and (3) presenting all discount rate combinations of other costs and benefits with SC-CO$_2$ estimates. As discussed above ("Approaches to Discounting and Their Applications"), combining estimates of costs and benefits when only some categories have an intergenerational component raises challenges that have yet to be resolved.

OTHER DISCOUNTING ISSUES

Time Consistency and Uncertain Discount Rates

One objection frequently made to the use of a declining discount rate is that it may lead to problems of time inconsistency. Time inconsistency refers to a change in expected net benefits due solely to the passing of time. Consider what happens when the expected present value of damages associated with emitting 1 metric ton of CO$_2$ in 2035 is evaluated in 2015 (e.g., for a regulation implemented in 2015 that affects emissions in 2035). Damages occurring in 2055 from those 2035 emissions are discounted back to 2035 using a certainty-equivalent rate of 2.85 percent (see Table 6-2 above). Now imagine the damages associated with emitting 1 metric ton of carbon in 2035 are recomputed in 2035 and the discount schedule in Table 6-2 has not changed. That is, the first column of discount factor/rates is the same (1 and 3 percent) but applies to 2035. Now damages in 2055 will be discounted to 2035 using a higher certainty-equivalent rate of 2.9 percent. This occurs because in 2035, 2055 will be closer to the present. Although the changes are small, one can imagine larger effects over longer horizons that might affect the preference for one or another option.

This apparent inconsistency is not in fact inconsistent. The discount rate schedule shown in Table 6-2 corresponds to the distribution of future growth rates given information available in 2015. At present, no one knows what the distribution of future growth rates in 2035 will be; it may be different or the same as the distribution in 2015. Even if it turns out to be the same as the distribution in 2015, that realization is new information that was not available in 2015.

Regional Disaggregation and Discounting

The possibility of disaggregating damages by geographic region (see Chapter 5) raises the issue of whether region-specific discount rates might be used to discount damages. Two approaches can be taken to the issues of aggregating damages across regions and discounting. Damages can be aggregated across regions in a given year to yield global damages, and global damages can be discounted, as described above. Alternately, damages could be discounted to the present for each region and then aggregated. The discount rates in each region could be region-specific based on region-specific growth rates. Currently, the IWG employs the former approach.

Using region-specific discount rates requires values of δ and η for each region, as well as a distribution over the rate of growth of per capita consumption in each region. Treating future generations differently based on where they live—whether due to differing values of δ and η or to differing growth rates—suggests a need to treat current generations differently on the basis of where they live. This raises the issue of how such regional weights would be determined.[24] The current approach avoids this issue and, by applying the same discount rate to all countries, is in the spirit of OMB guidance, which calls for treating equally persons of different income levels at a given time, for the purposes of valuation.

---

[24]Note that in a simple formulation, η would be a basis for both how marginal utility declines with economic growth and how differences in levels of economic development are weighted. This need not be the case in general.

7

Long-Term Research Needs

In Chapters 3-5 the committee provides recommendations and conclusions for both near-term and long-term improvements to SC-CO$_2$ estimation. In this chapter, the conclusions that focus solely on long-term research needs as a guide for research in the relevant fields are repeated. The committee notes that neither the IWG nor any other single entity has responsibility for identifying and supporting research in these fields.[1] Thus, these conclusions about what is needed are intended for all interested researchers, institutions that support research, and policy makers. For each component of the SC-CO$_2$ analysis discussed below, research topics are listed in order of priority for developing updates to the SC-CO$_2$ framework.

The committee structured its work, conclusions, and recommendations around four components of analysis that are involved in estimating the SC-CO$_2$—socioeconomic and emissions projections, climate modeling, estimation of climate damages, and discounting net monetary damages—which are identified as modules. Each module comprises conceptual formulations and theory, computer models and other analytical frameworks, and each is supported by its own specialized disciplinary

---

[1]Recognizing that the IWG is itself not a research funding agency, we encourage the IWG to communicate these research priorities to the key research programs of its member agencies, as well as the U.S. Global Change Research Program (USGCRP), the National Science Foundation, and other funding agencies of the USGCRP, and other relevant research and research funding institutions both within and outside of government.

expertise. Among the committee's research recommendations, the highest priority is placed on research relevant to the climate damages module followed by the socioeconomic module.

Estimation of the SC-CO_2 involves the integration of these four modules, while taking account, when possible, of the feedbacks and interactions among them. Research in climate impacts and damages, integrated assessment, economics, and Earth system modeling could explore interactions in and feedbacks among the components of SC-CO_2 estimation that go beyond the potential feedback of climate impacts on socioeconomic projections, or that may even suggest changes in the structure of the four-module system. In combination, these physical and economic changes might yield regional and global interactions large enough to affect the overall operation of the economic or climate system in ways that have not yet been considered. It is not clear in what ways the structure and implementation of the SC-CO_2 estimates might be refined to capture changes in understanding, but accounting for new knowledge will be important to future updates of the SC-CO_2 estimates.

In addition, three of the committee's recommended modules—socioeconomic, climate, and damages—require advances in the models that contribute to their component of SC-CO_2 estimation. For the fourth module, on discounting, the committee's recommendations rely on applying existing techniques to the SC-CO_2 estimation process, so we do not offer any specific research needs in this area. However, the committee does not mean to imply that the issue of long-term discounting would not benefit from further research.

INTERACTIONS

CONCLUSION 2-3 Research to identify and explore the magnitude of various interactions and feedbacks within the human-climate system, which are relationships not currently well represented in the SC-CO_2 estimation framework, will be an important input to longer-term enhancements in the SC-CO_2 estimation framework. Areas of research that are likely to yield particular benefits include:

- **Exploration of methods for representing feedbacks among systems and interactions within them, such as:**
  - **feedbacks between climate, physical impacts, economic damages, and socioeconomic projections, and**
  - **interactions between types of impacts or economic damages within and across regions of the world.**

- Assessment of the relative importance of specific feedbacks and interactions in the estimation of the SC-CO$_2$, perhaps using an existing detailed structure model of the world economy.
- Assessment of existing analyses that integrate socioeconomic, climate, and damage components to assess their suitability for use in estimating the SC-CO$_2$, particularly with respect to feedbacks and interactions, while recognizing the computational requirements for such analyses.

SOCIOECONOMIC AND EMISSIONS PROJECTIONS

The IWG process is committed to updating the SC-CO$_2$ estimates as the science and economic understanding of climate change and its impacts on society improve over time. There are therefore many advantages to encouraging research that supports the construction of a dedicated socioeconomic projection framework and, considering its unique objectives, a detailed-structure economic model, as recommended in Chapter 3.

CONCLUSION 3-1 Research on key elements of long-term economic and energy models and their inputs, focused on the particular needs of socioeconomic projections in SC-CO$_2$ estimation, would contribute to the design and implementation of a new socioeconomic module. Interrelated areas of research that could yield particular benefits include the following, in rough order of priority:

- Development of a socioeconomic module to support damage estimates that depend on interactions within the human-climate system (e.g., among energy, water, and agriculture, and between urban emissions and air pollution).
- Use of econometric and other methods to construct long-run projections of population and gross domestic product (GDP) and their uncertainties.
- Quantification of the magnitude of feedbacks of climate outputs and various measures of damages (e.g., on consumption, productivity, and capital stocks) on socioeconomic projections, based in part on existing detailed-structure models.
- Development of detailed-structure economic models suited to projections that are consistent over very long time horizons, in which functional form and levels of regional and sectoral

detail in inputs and outputs may differ between the nearer term (e.g., to 2100) and the more distant future.

- Development of probability distributions of uncertain parameters used in detailed-structure models, with a particular focus on the differences among developed, transitional and low-income economies. Examples of uncertain parameters include key elasticities of substitution (e.g., between labor and capital inputs to production, between energy and nonenergy demand, and among fuels in total energy use), energy technology costs and rates of technology penetration, and rates of capital turnover.

EARTH SYSTEM MODELING

In this area, the committee's identified research needs cover both the near term and the long term.

CONCLUSION 4-5 Research focused on improving the representation of the Earth system in the context of coupled climate-economic analyses would improve the reliability of estimates of the SC-CO_2. In the near term, research in six areas could yield benefits for SC-CO_2 estimation:

- coordinated research to reduce uncertainty in estimates of the capacity of the land and ocean to absorb and store carbon, especially in the first century after a pulse release, applied to a range of scenarios of future atmospheric composition and temperature;
- coordinated Earth system model experiments injecting identical pulses of CO_2 and other greenhouse gases in a range of scenarios of future atmospheric composition and temperature;
- the development of simple, probabilistic sea level rise models that incorporate the emerging science on ice sheet stability and that can be linked to simple Earth system models;
- systematic assessments of the dependence of patterns of regional climate change on spatial patterns of forcing, the relationship between regional climate extremes and global mean temperature, the temporal evolution of patterns under conditions of stable or decreasing forcing, and nonlinearities in the relationship between global means and regional variables;

- systematic assessments of nonlinear responses to forcing in Earth system models and investigations into evidence for such responses in the geological record; and
- the development of simple Earth system models that incorporate nonlinear responses to forcing and assessments of the effects of such nonlinear responses on SC-CO$_2$ estimation.

In the longer term, more comprehensive climate models could be incorporated into the SC-CO$_2$ estimation framework. However, the major focus of current model research is on increasing resolution and comprehensiveness, rather than on expanding the ability of comprehensive models to be used for risk analysis. SC-CO$_2$ estimation would be advanced by an expanded focus on probabilistic methods that use comprehensive Earth system models, including the use of comprehensive models to represent low-probability, high-consequence states of the world, as well as the use of decision support science approaches to identify and evaluate key decision-relevant uncertainties in Earth system models.

CLIMATE DAMAGE ESTIMATION

Finally, the committee outlines in Chapter 5 a set of desirable characteristics of a damages module that could be developed in the long term and would improve the reliability of estimates of the SC-CO$_2$. The committee's conclusions cover the research tasks that would support the development of such a module.

CONCLUSION 5-1 An expansion of research on climate damage estimation is needed and would improve the reliability of estimates of the SC-CO$_2$.

- In the near term, initial steps that could be undertaken include:
  - a comprehensive review of the literature on climate impacts and damage estimation, the evaluation of adaptation responses, and regional and sectoral interactions, as well as feedbacks among the damage, socioeconomic, and climate modules; and
  - a comparison of methods for estimating damages, including characterizations of their differences, synergies, uncertainties, and treatment of adaptation.
- In the medium to long term, several research priorities could yield particular benefits for SC-CO$_2$ estimation:

- physical, structural economic, and empirical estimation of climate impact relationships for regions and sectors not currently covered in the peer-reviewed literature;
- structural and empirical studies of the efficacy and costs of adaptation;
- calibration of damage functions using empirical and structural models operating at sufficiently high temporal and spatial resolution to capture relevant dynamics;
- the development of systematic frameworks for translating estimates of impacts into welfare costs; and
- empirical observation-based and structural modeling studies of interregional and intersectoral interactions of impacts, as well as of feedbacks among damages, socioeconomic factors, and emissions.

- In the long term, research priorities that could yield particular benefits for SC-CO_2 estimation would include omitted critical thresholds in natural and socioeconomic systems:
  - development of simple Earth system model or full complexity Earth system model scenarios in which potential critical thresholds of tipping elements (e.g., Atlantic meridional overturning circulation, monsoonal circulation patterns, sea ice, polar ice sheets) are crossed, and the use of the physical changes in these scenarios to drive models that assess impacts and damages;
  - empirical observation-based and structural modeling studies of the potential for climate change to drive the crossing of critical thresholds in socioeconomic systems and of their ensuing damages; and
  - expert elicitation studies of the likelihood of different tipping element scenarios, in order to allow tipping elements and their critical thresholds to be represented probabilistically in the SC-CO_2 framework.

Overall, the committee's long-term recommendations on an integrated approach to estimating the SC-CO_2, as well as the socioeconomic, climate, and damages modules, requires a significant advance in the scientific literature. It is important that the IWG continue to engage with the scientific community to produce the research identified above. As noted in the committee's recommendation for a regularized updating process in Chapter 2 (Recommendation 2-4), research, scientific advances, and peer review are central elements to improving the reliability and transparency of the SC-CO_2 estimates.

References

Ackerman, F., Stanton, E.A., and Bueno, R. (2010). Fat tails, exponents, extreme uncertainty: Simulating catastrophe in DICE. *Ecological Economics, 69*(8), 1657-1665.

Ackerman, F., Stanton, E.A., and Bueno, R. (2013). Epstein–Zin utility in DICE: Is risk aversion irrelevant to climate policy? *Environmental and Resource Economics, 56,* 73.

Agnew, M., Schrattenholzer, L., and Voss, A. (1978). *User's Guide for the MESSAGE Computer Program.* RM-78-026. Laxenburg, Austria: International Institute for Applied Systems Analysis.

Alley, R.B., Marotzke, J., Nordhaus, W.D., Overpeck, J.T., Peteet, D.M., Pielke, R.A., Pierrehumbert, R.T., Rhines, P.B., Stocker, T.F., Talley, L.D., and Wallace, J.M. (2003). Abrupt climate change. *Science, 299*(5615), 2005-2010.

Altizer, S., Ostfeld, R.S., Johnson, P.T., Kutz, S., and Harvell, C.D. (2013). Climate change and infectious diseases: From evidence to a predictive framework. *Science, 341*(6145), 514-519.

Andrews, T., Gregory, J.M., Webb, M.J., and Taylor, K.E. (2012). Forcing, feedbacks and climate sensitivity in CMIP5 coupled atmosphere-ocean climate models. *Geophysical Research Letters, 39*(9), 1-7.

Anthoff, D., and Tol, R.S.J. (2014). *FUND—Climate Framework for Uncertainty, Negotiation and Distribution.* Version 3.8. Available: http://www.fund-model.org/versions [October 2016].

Anthoff, D., Nicholls, R.J., Tol, R.S.J., and Vafeidis, A.T. (2006). *Global and Regional Exposure to Large Rises in Sea-Level: A Sensitivity Analysis.* Working Paper 96. Norwich, U.K.: Tyndall Centre for Climate Change Research.

Arent, D.J., Tol, R.S.J., Faust, E., Hella, J.P., Kumar, S., Strzepek, K.M., Tóth, F.L., and Yan, D. (2014). Key economic sectors and services—supplementary material. In C.B. Field, V.R. Barros, D.J. Dokken, K.J. Mach, M.D. Mastrandrea, T.E. Bilir, M. Chatterjee, K.L. Ebi, Y.O. Estrada, R.C. Genova, B. Girma, E.S. Kissel, A.N. Levy, S. MacCracken, P.R. Mastrandrea, and L.L. White (Eds.), *Climate Change 2014: Impacts, Adaptation, and Vulnerability. Part A: Global and Sectoral Aspects. Contribution of Working Group II to the Fifth Assessment Report of the Intergovernmental Panel on Climate Change.* Available: http://www.ipcc.ch/report/ar5/wg2 and www.ipcc.ch [January 2017].

Armour, K.C., Bitz, C.M., and Roe, G.H. (2012). Time-varying climate sensitivity from regional feedbacks. *Journal of Climate, 26*(13), 4518-4534.

Arora, V.K., Boer, G.J., Friedlingstein, P., Eby, M., Jones, C.D., Christian, J.R., Bonan, G., Bopp, L., Brovkin, V., Cadule, P., Hajima, T., Ilyina, T., Lindsay, K., Tjiputra, J.F., and Wu, T. (2013). Carbon-concentration and carbon-climate feedbacks in CMIP5 earth system models. *Journal of Climate, 26*(15), 5289-5314.

Arrow, K.J., Cline, W.R., Maler, K.G., Munasinghe, M., Squitieri, R., and Stiglitz, J.E. (1996). Chapter 4: Intertemporal Equity, Discounting, and Economic Efficiency. In *Climate Change 1995: Economic and Social Dimensions of Climate Change Contribution of Working Group III to the Second Assessment Report of the Intergovernmental Panel on Climate Change* (pp. 125-144). Available: https://www.ipcc.ch/ipccreports/sar/wg_III/ipcc_sar_wg_III_full_report.pdf [January 2017].

Arrow, K.J., Cropper, M.L., Gollier, C., Groom, B., Heal, G.M, Newell, R.G., Nordhaus, W.D., Pindyck, R.S., Pizer, W.A., Portney, P.R., Sterner, T., Tol, R.S.J, and Weitzman, M.L. (2013). Determining benefits and costs for future generations. *Science, 341*(6144), 349-350.

Arrow, K.J., Cropper, M.L., Gollier, C., Groom, B., Heal, G.M., Newell, R.G., Nordhaus, W.D., Pindyck, R.S., Pizer, W.A., Portney, P.R., Sterner, T., Tol, R.S.J., and Weitzman, M.L. (2014). Should governments use a declining discount rate in project analysis? *Review of Environmental Economics and Policy, 8*(2), 145-163.

Atkinson, G., Dietz, S., Helgeson, J., Hepburn, C., and Sælend, H. (2009). Siblings, not triplets: Social preferences for risk, inequality and time in discounting climate change. *Economics, 3*(26), 1-28.

Auffhammer, M., and Aroonruengsawat, A. (2012). *Hotspots of Climate-Driven Increases in Energy Demand: A Simulation Exercise Based on Household Level Billing Data for California.* Available: http://www.energy.ca.gov/2012publications/CEC-500-2012-021/CEC-500-2012-021.pdf [October 2016].

Auffhammer, M., Ramanathan, V., and Vincent J. (2012). Observation-based evidence that climate change has reduced Indian rice harvests. (2012). *Climatic Change, 111*(2), 411-424.

Auffhammer, M., and Mansur, E.T. (2014). Measuring climatic impacts on energy consumption: A review of the empirical literature. *Energy Economics, 46*, 522-530.

Ayres, R., and Walter, J. (1991). The greenhouse effect: Damages, costs and abatement. *Environmental & Resource Economics* 1(3), 237-270.

Azar, C., and Lindgren, K. (2003). Catastrophic events and stochastic cost-benefit analysis of climate change. *Climatic Change, 56*(3), 245-255.

Baldos, U.L.C., and Hertel, T.W. (2014). Global food security in 2050: The role of agricultural productivity and climate change. *Australian Journal of Agricultural and Resource Economics, 58*(4), 554-570.

Barreca, A., Clay, K., Deschenes, O., Greenstone, M., and Shapiro, J.S. (2013). *Adapting to Climate Change: The Remarkable Decline in the U.S. Temperature-Mortality Relationship over the 20th Century.* Working Paper No. 18692. Cambridge, MA: National Bureau of Economic Research.

Barreca, A.I., Clay, K., Deschenes, O., Greenstone, M., and Shapiro, J.S. (2015). *Adapting to Climate Change: The Remarkable Decline in the U.S. Temperature-Mortality Relationship over the 20th Century.* IZA discussion papers, No. 8915. Available: http://www.nber.org/papers/w18692 [April 2017].

Barro, R.J., and Ursúa, J.F. (2008a). Consumption disasters in the twentieth century. *The American Economic Review, 98*(2), 58-63.

Barro, R.J., and Ursúa, J.F. (2008b). Macroeconomic crises since 1870. *Brookings Papers on Economic Activity, 39*(1), 255-350.

Barton, A., Hales, B., Waldbusser, G.G., Langdon, C., and Feely, R.A. (2012). The Pacific oyster, Crassostrea gigas, shows negative correlation to naturally elevated carbon dioxide levels: Implications for near-term ocean acidification effects. *Limnology and Oceanography, 57*(3), 698-710.

Basten, S., Lutz, W., and Sherbov, S. (2013). Very long range global population scenarios to 2300 and the implications of sustained low fertility. *Demographic Research, 28*(39), 1145-1166.

Bates, N.R. (2007). Interannual variability of the oceanic CO_2 sink in the subtropical gyre of the North Atlantic Ocean over the last 2 decades. *Journal of Geophysical Research, 112*(C9), 1-26.

Bijlsma, L., Ehler, C.N., Klein, R.J.T., Kulshrestha, S.M., McLean, R.F., Mimura, N., Nicholls, R.J., Nurse, L.A., Perez Nieto, H., Stakhiv, E.Z., Turner, R.K., and Warrick, R.A. (1996). Coastal zones and small islands. In R.T. Watson, M.C. Zinyowera, and R.H. Moss (Eds.), *Climate Change 1995: Impacts, Adaptations and Mitigation of Climate Change: Scientific-Technical Analyses—Contribution of Working Group II to the Second Assessment Report of the Intergovernmental Panel on Climate Change* (1st Edition) (pp. 289-324). Cambridge, U.K.: Cambridge University Press,

Blanford, G., Merrick, J., Richels, R., and Steven, R. (2014). Trade-offs between mitigation costs and temperature change. *Climatic Change, 123*(3-4), 527-541.

Bloch-Johnson, J., Pierrehumbert, R.T., and Abbot, D.S. (2015). Feedback temperature dependence determines the risk of high warming. *Geophysical Research Letters, 42*(12), 4973-4980.

Bopp, L., Resplandy, L., Orr, J.C., Doney, S.C., Dunne, J.P., Gehlen, M., Halloran, P., Heinze, C., Ilyina, T., Séférian, R., Tjiputra, J., and Vichi, M. (2013). Multiple stressors of ocean ecosystems in the 21st century: Projections with CMIP5 models, *Biogeosciences, 10*, 6225-6245.

Bosello, F., Roson, R., and Tol, R.S.J. (2007). Economy-wide estimates of the implications of climate change: Sea level rise. *Environmental and Resource Economics, 37*(3), 549-571.

Bosello, F., Nicholls, R.J., Richards, J., Roson, R., and Tol, R.S.J. (2012). Economic impacts of climate change in Europe: Sea-level rise. *Climatic Change, 112*(1), 63-81.

Bosetti, V., Carraro, C., Galeotti, M., Massetti, E., and Tavoni, M. (2006). WITCH: A World Induced Technical Change Hybrid Model. *The Energy Journal, 27*(SI2), 13-37.

Bosetti, V., Marangoni, G., Borgonovo, E., Diaz Anadon, L., Barron, R., McJeon, H., Politis, S., and Friley, P. (2015). Sensitivity to energy technology costs: A multi-model comparison analysis. *Energy Policy, 80*, 244-263.

Bradford, D.F. (1975). Constraints on government investment opportunities and the choice of discount rate, *The American Economic Review, 65*(5), 887-899. Available: http://www.jstor.org/stable/1806627?seq=1#page_scan_tab_contents [January 2017]

Brander, L.M., Florax, R.J.G.M., and Vermaat, J.E. (2006). The empirics of wetland valuation: A comprehensive summary and a meta-analysis of the literature. *Environmental and Resource Economics, 33*(2), 223-250.

Burke, M., Hsiang, S., Miguel, E. (2015). Global non-linear effect of temperature on economic production, *Nature, 527*, 235-239.

Butler, E.E., and Huybers, P. (2013). Adaptation of U.S. maize to temperature variations. *Nature Climate Change, 3*(1), 68-72.

Caballero, R., and Huber, M. (2013). State-dependent climate sensitivity in past warm climates and its implications for future climate projections. *Proceedings of the National Academy of Sciences of the United States of America, 110*(35), 14162-14167.

Cai, Y., Lenton, T.M., and Lontzek, T.S. (2016). Risk of multiple interacting tipping points should encourage rapid CO_2 emission reduction. *Nature Climate Change, 6*(5), 520-525.

Caldeira, K., and Myhrvold, N.P. (2013). Projections of the pace of warming following an abrupt increase in atmospheric carbon dioxide concentration. *Environmental Research Letters, 8*(3), 034039.

Caldwell, P., and Bretherton, C.S. (2009). Response of a subtropical stratocumulus-capped mixed layer to climate and aerosol changes. *Journal of Climate, 22*, 20-38.

Calvin, K., Wise, M., Clarke, L., Edmonds, J., Kyle, G., Luckow, P., and Thomson, A. (2013). Implications of simultaneously mitigating and adapting to climate change: Initial experiments using GCAM. *Climatic Change, 117*(3), 545-560.

Caminade, C., Kovats, S., Rocklov, J., Tompkins, A.M., Morse, A.P., Colón-González, F.J., Stenlund, H., Martens, P., and Lloyd, S.J. (2014). Impact of climate change on global malaria distribution. *Proceedings of the National Academy of Sciences of the United States of America, 111*(9), 3286-3291.

Carleton, T.A., and Hsiang, S.M. (2016). Social and economic impacts of climate. *Science, 353*(6304).

Carlos, C.M.J., Luc, F., Antonio, S.R., Carlo, L., Miles, P., Frank, R., Francoise, N., Hande, D., Máté, R., Alessandro, D., Marcello, D., Kumar, S.A., Davide, F., Stefan, N., Shailesh, S., Pavel, C., Mihaly, H., Benjamin, V.D., Salvador, B., Nicolás, I.R.J., Giovanni, F., Felipe, R.M.R., Alessandra, B., Paul, D., Andrea, C., Giorgio, L., Jesus, S.-M.-A., Daniele, D.R., Giovanni, C., Ignacio, B.C.J., Daniele, P., Jonathan, P., Bert, S., Tamas, R., Claudia, B., Ine, V., Filipe, B.E.S., and Dolores, I.R. (2014). *Climate Impacts in Europe. The JRC PESETA II Project.* Report EUR 26586EN. Luxembourg: European Union. Available: http://publications.jrc.ec.europa.eu/repository/handle/JRC87011 [September 2016].

Castruccio, S., McInerney, D.J., Stein, M.L., Crouch, F.L., Jacob, R.L., and Moyer, E.J. (2014). Statistical emulation of climate model projections based on precomputed GCM runs. *Journal of Climate, 27*(5), 1829-1844.

Chen, Y.-H.H., Paltsev, S., Reilly, J.M., Morris, J.F., and Babiker, M.H. (2015). *The MIT EPPA6 Model: Economic Growth, Energy Use, and Food Consumption.* Report 278. Cambridge, MA: MIT Joint Program on the Science and Policy of Global Change.

Chen, Y.-H.H., Paltsev, S., Reilly, J., Morris, J.F., and Babiker, M.H. (2016). Long-term economic modeling for climate change assessment. *Economic Modeling, 52*(Part B), 867-883.

Church, J.A., Clark, P.U., Cazenave, A., Gregory, J.M., Jevrejeva, S., Levermann, A., Merrifield, M.A., Milne, G.A., Nerem, R.S., Nunn, P.D., Payne, A.J., Pfeffer, W.T., Stammer, D., and Unnikrishnan, A.S. (2013). Sea level change. In T.F. Stocker, D. Qin, G.-K. Plattner, M. Tignor, S.K. Allen, J. Boschung, A. Nauels, Y. Xia, V. Bex, and P.M. Midgley (Eds.), *Climate Change 2013: The Physical Science Basis. Contribution of Working Group I to the Fifth Assessment Report of the Intergovernmental Panel on Climate Change* (Ch. 13) (pp. 1137-1216). Cambridge, U.K. and New York: Cambridge University Press.

Chuwah, C., van Noije, T., van Vuuren, D.P., Hazeleger, W., Strunk, A., Deetman, S., Beltran, A.M., and van Vliet, J. (2013). Implications of alternative assumptions regarding future air pollution control in scenarios similar to the Representative Concentration Pathways. *Atmospheric Environment, 79*, 787-801.

Ciais, P., Sabine, C., Bala, G., Bopp, L., Brovkin, V., Canadell, J., Chhabra, A., DeFries, R., Galloway, J., Heimann, M., Jones, C., Le Quéré, C., Myneni, R.B., Piao, S., and Thornton, P. (2013). Carbon and other biogeochemical cycles. In T.F. Stocker, D. Qin, G.-K. Plattner, M. Tignor, S.K. Allen, J. Boschung, A. Nauels, Y. Xia, V. Bex, and P.M. Midgley (Eds.), *Climate Change 2013: The Physical Science Basis. Contribution of Working Group I to the Fifth Assessment Report of the Intergovernmental Panel on Climate Change* (Ch. 6) (pp. 465-570). Cambridge, U.K. and New York: Cambridge University Press.

Ciscar, J.-C., Iglesias, A., Feyen, L., Szabo, L., Van Regemorter, D., Amelung, B., Nicholls, R., Watkiss, P., Christensen, O., Dankers, R., Garrote, L., Goodess, C., Hunt, A., Moreno, A., Richards, J., and Soria, A. (2011). Physical and economic consequences of climate change in Europe. *Proceedings of the National Academy of Sciences of the United States of America, 108*(7), 2678-2683.

Ciscar, J.-C., Feyen, L., Soria, A., Lavalle, C., Raes, F., Perry, M., Nemry, F., Demirel, H., Rozsai, M., Dosio, A., Donatelli, M., Srivastava, A., Fumagalli, D., Niemeyer, S., Shrestha, S., Ciaian, P., Himics, M., van Doorslaer, B., Barrios, S., Ibáñez, N., Forzieri, G., Rojas, R., Bianchi, A., Dowling, P., Camia, A., Libertà, G., San Miguel, J., de Rigo, D., Caudullo, G., Barredo, J.I., Paci, D., Pycroft, J., Saveyn, B., van Regemorter, D., Revesz, T., Vandyck, T., Vrontisi, Z., Baranzelli, C., Vandecasteele, I., Batista e Silva, F., and Ibarreta, D. (2014). *Climate Impacts in Europe: The JRC PESETA II Project.* Luxembourg: European Union.

Clarke, L., Edmonds, J., Krey, V., Richels, R., Rose, S., and Tavoni, M. (2009). International policy architectures: Overview of the EMF-22 International Scenarios. *Energy Economics, 31*(Suppl. 2), S64-S81.

Clemen, R.T., and Winkler, R.L. (1985). Limits for the precision and value of information from dependent sources. *Operations Research, 33*(2), 427-442.

Collier, P., Elliott V.L., Hegre H., Hoeffler A., Reynal-Querol M., and Sambanis, N. (2003). *Breaking the Conflict Trap: Civil War and Development Policy.* Washington, DC: The World Bank.

Collins, M., Knutti, R., Arblaster, J., Dufresne, J.-L., Fichefet, T., Friedlingstein, P., Gao, X., Gutowski, W.J., Johns, T., Krinner, G., Shongwe, M., Tebaldi, C., Weaver, A.J., and Wehner, M. (2013). Long-term climate change: Projections, commitments and irreversibility. In T.F. Stocker, D. Qin, G.-K. Plattner, M. Tignor, S.K. Allen, J. Boschung, A. Nauels, Y. Xia, V. Bex, and P.M. Midgley (Eds.), *Climate Change 2013: The Physical Science Basis. Contribution of Working Group I to the Fifth Assessment Report of the Intergovernmental Panel on Climate Change* (Ch. 12) (pp. 1029-1136). Cambridge, U.K. and New York: Cambridge University Press.

Cooley, S.R., and Doney, S.C. (2009). Anticipating ocean acidification's economic consequences for commercial fisheries. *Environmental Research Letters, 4*(2), 1-8.

Cooley, S.R., Rheuban, J.E., Hart, D.R., Luu, V., Glover, D.M., Hare, J.A., and Doney, S.C. (2015). An integrated assessment model for helping the United States sea scallop (*Placopecten magellanicus*) fishery plan ahead for ocean acidification and warming. *PLoS One, 10*(5), 1-27.

Cropper, M.L., Freeman, M.C., Groom, B., and Pizer, W.A. (2014). Declining discount rates. *The American Economic Review, 104*(5), 538-543.

Crucifix, M. (2006). Does the last glacial maximum constrain climate sensitivity? *Geophysical Research Letters, 33*(18), 1-5.

Daioglou, V., Wicke, B., Faaijand, A.P.C., and van Vuuren, D.P. (2014). Competing uses of biomass for energy and chemicals: Implications for long-term global CO_2 mitigation potential. *GCB Bioenergy, 7*(6), 1321-1334.

Darwin, R., Tsigas, M., Lewandrowski, J., and Raneses, A. (1995). *World Agriculture and Climate Change.* AER-703. Available: https://www.ers.usda.gov/webdocs/publications/aer703/32471_aer703_002.pdf?v=41304 [January 2017].

Davie, J.C.S., Falloon, P.D., Kahana, R., Dankers, R., Betts, R., Portmann, F.T., Wisser, D., Clark, D.B., Ito, A., Masaki, Y., Nishina, K., Fekete, B., Tessler, Z., Wada, Y., Liu, X., Tang, Q., Hagemann, S., Stacke, T., Pavlick, R., Schaphoff, S., Gosling, S.N., Franssen, W., and Arnell, N. (2013). Comparing projections of future changes in runoff from hydrological and biome models in ISI-MIP. *Earth System Dynamics*, 4(2), 359-374.

DeConto, R.M., and Pollard, D. (2016). Contribution of Antarctica to past and future sea-level rise. *Nature 531*, 591-597.

Dell, M., Jones, B.F., and Olken, B.A. (2012). Temperature shocks and economic growth: Evidence from the last half century. *American Economic Journal: Macroeconomics*, 4(3), 66-95.

Dell, M., Jones, B.F., and Olken, B.A. (2014). What do we learn from the weather? The new climate-economy literature. *Journal of Economic Literature*, 52(3), 740-798.

Deschenes, O. (2014). Temperature, human health, and adaptation: A review of the empirical literature. *Energy Economics, 46*, 606-619.

Diaz, D.B. (2016). Estimating global damages from sea level rise with the Coastal Impact and Adaptation Model (CIAM). *Climatic Change*, 137(1), 143-156.

Dickson, A.G. (1981). An exact definition of total alkalinity and a procedure for the estimation of alkalinity and total inorganic carbon from titration data. *Deep Sea Research Part A Oceanographic Research Papers, 28*(6), 609-623.

Diffenbaugh, N.S., Hertel, T.W., Scherer, M., and Verma, M. (2012). Response of corn markets to climate volatility under alternative energy futures. *Nature Climate Change, 2*, 514-518.

Doney, S., Balch, W., Fabry, V., and Feely, R. (2009). Ocean acidification: A critical emerging problem for the ocean sciences. *Oceanography, 22*(4), 16-25.

Dore, J.E., Lukas, R., Sadler, D.W., Church, M.J., and Karl, D.M. (2009). Physical and biogeochemical modulation of ocean acidification in the central North Pacific. *Proceedings of the National Academy of Sciences of the United States of America, 106*(30), 12235-12240.

Downing, T.E., Greener, R.A., and Eyre, N. (1995). *The Economic Impacts of Climate Change: Assessment of Fossil Fuel Cycles for the ExternE Project.* Oxford and Lonsdale, U.K.: Environmental Change Unit and Eyre Energy Environment.

Downing, T.E., Eyre, N., Greener, R., and Blackwell, D. (1996). *Full Fuel Cycle Study: Evaluation of the Global Warming Externality for Fossil Fuel Cycles with and without CO_2 Abatement and for Two Reference Scenarios, Environmental Change Unit.* Oxford, U.K.: University of Oxford.

Drijfhout, S. (2015). Competition between global warming and an abrupt collapse of the AMOC in Earth's energy imbalance. *Scientific Reports, 5, 14877*.

Edmonds, J., and Reilly, J. (1983). Global energy and CO_2 to the year 2050. *The Energy Journal*, 4(3), 21-47.

Epstein, L.G., and Zin, S.E. (1989). Substitution, risk aversion, and the temporal behavior of consumption and asset returns: A theoretical framework. *Econometrica, 57*, 937-969.

Epstein, L.G., and Zin, S.E. (1991). Substitution, risk aversion, and the temporal behavior of consumption and asset returns: An empirical analysis. *Journal of Political Economy, 99*, 263-286.

Fabry, V.J., Seibel, B.A., Feely, R.A., and Orr, J.C. (2008). Impacts of ocean acidification on marine fauna and ecosystem processes. *ICES Journal of Marine Science, 65*(3), 414-432.

Fankhauser, S. (1994). The social costs of greenhouse gas emissions: An expected value approach. *The Energy Journal, 15*(2), 157-184.

Fischer, G., Frohberg, K., Parry, M.L., and Rosenzweig, C. (1996). Impacts of potential climate change on global and regional food production and vulnerability. In T.E. Downing (Ed.), *Climate Change and World Food Security* (pp. 115-159). Berlin, Germany: Springer-Verlag.

Flato, G., Marotzke, J., Abiodun, B., Braconnot, P., Chou, S.C., Collins, W., Cox, P., Driouech, F., Emori, S., Eyring, V., Forest, C., Glecker, P., Guilyardi, E., Jakob, C., Kattsov, V., Reason, C., and Rummukainen, M. (2013). Evaluation of climate models. In T.F. Stocker, D. Qin, G.-K. Plattner, M. Tignor, S.K. Allen, J. Boschung, A. Nauels, Y. Xia, V. Bex, and P.M. Midgley (Eds.), *Climate Change 2013: The Physical Science Basis. Contribution of Working Group I to the Fifth Assessment Report of the Intergovernmental Panel on Climate Change* (Ch. 9) (pp. 741-866). Cambridge, U.K. and New York: Cambridge University Press.

Fowler, H.J., Blenkinsop, S., and Tebaldi, C. (2007). Linking climate change modelling to impacts studies: Recent advances in downscaling techniques for hydrological modelling. *International Journal of Climatology, 27*, 1547-1578.

Frame, D.J., Booth, B.B.B., Kettleborough, J.A., Stainforth, D.A., Gregory, J.M., Collins, M., and Allen, M.R. (2005). Constraining climate forecasts: The role of prior assumptions. *Geophysical Research Letters, 32*(9), 1-5.

Frame, D.J., Stone, D.A., Stott, P.A., and Allen, M.R. (2006). Alternatives to stabilization scenarios. *Geophysical Research Letters, 33*(14), 1-4.

Frieler, K., Meinshausen, M., Mengel, M., Braun, N., and Hare, W. (2012). A scaling approach to probabilistic assessment of regional climate change. *Journal of Climate, 25*, 3117-3144.

Gattuso, J.-P., Magnan, A., Billé, R., Cheung, W.W.L., Howes, E.L., Joos, F., Allemand, D., Bopp, L., Cooley, S.R., Eakin, C.M., Hoegh-Guldberg, O., Kelly, R.P., Pörtner, H.-O., Rogers, A.D., Baxter, J.M., Laffoley, D., Osborn, D., Rankovic, A., Rochette, J., Sumaila, U.R., Treyer, S., and Turley, C. (2015). Contrasting futures for ocean and society from different anthropogenic CO_2 emissions scenarios. *Science, 349*(6243):aac4722.

Geoffroy, O., Saint-Martin, D., Bellon, G., Voldoire, A., Olivié, D.J.L., and Tytéca, S. (2013). Transient climate response in a two-layer energy-balance model. Part II: Representation of the efficacy of deep-ocean heat uptake and validation for CMIP5 AOGCMs. *Journal of Climate, 26*(6), 1859-1876.

Gerland, P., Raftery, A.E., Šev íková, H., Li, N., Gu, D., Spoorenberg, T., Alkema, L, Fosdick, B.K., Chunn, J., Lalic, N., Bay, G., Buettner, T., Heilig, G.K., and Wilmoth, J. (2014). World population stabilization unlikely this century. *Science, 346*(6206), 234-237.

Giglio, S., Maggiori, M., and Stroebel, J. (2015). Very long-run discount rates. *Quarterly Journal of Economics, 130*(1), 1-53.

Gillett, N.P., Arora, V.K., Matthews, D., and Allen, M.R. (2013). Constraining the ratio of global warming to cumulative CO_2 emissions using CMIP5 simulations. *Journal of Climate, 26*, 6844-6858.

Gillingham, K., Nordhaus, W., Anthoff, D., Blanford, G., Bosetti, V., Christensen, P., McJeon, H., Reilly, J., and Sztorc, P. (2015). *Modeling Uncertainty in Climate Change: A Multi-Model Comparison.* Cowles Foundation Discussion Paper No. 2022. Available: http://cowles.yale.edu/sites/default/files/files/pub/d20/d2022.pdf [October 2016].

Gollier, C. (2002). Time horizon and the discount rate. *Journal of Economic Theory, 107*(2), 463-473.

Gollier, C. (2008). Discounting with fat-tailed economic growth. *Journal of Risk and Uncertainty, 37*(2), 171-186.

Gollier, C. (2012). *Pricing the Planet's Future: The Economics of Discounting in an Uncertain World.* Princeton, NJ: Princeton University Press.

Gollier, C. (2014). Discounting and growth. *American Economic Review, 104*(5):534-537.

Gollier, C., and Hammitt, J. (2014). The long-run discount rate controversy. *Annual Review of Resource Economics, 6*, 273-295.

Graff Zivin, J., and Neidell, M. (2014). Temperature and the allocation of time: Implications for climate change. *Journal of Labor Economics, 32*(1), 1-26.

Gregory, J. M. (2000). Vertical heat transports in the ocean and their effect on time-dependent climate change. *Climate Dynamics, 16*(7), 501-515.

Gregory, J.M., and Forster, P.M. (2008). Transient climate response estimated from radiative forcing and observed temperature change. *Journal of Geophysical Research: Atmospheres*, *113*(D23).

Gregory, J.M., Jones, C.D., Cadule, P., and Friedlingstein, P. (2009). Quantifying carbon cycle feedbacks. *Journal of Climate*, *22*(19), 5232-5250.

Grinsted, A., Moore, J.C., and Jevrejeva, S. (2009). Reconstructing sea level from paleo and projected temperatures 200 to 2100 AD. *Climate Dynamics*, *34*(4), 461-472.

Grogan, D.S., Zhang, F., Prusevich, A., Lammers, R.B., Wisser, D., Glidden, S., Li, C., and Frolking, S. (2015). Quantifying the link between crop production and mined groundwater irrigation in China. *Science of the Total Environment*, *511*, 161-175.

Hakuba, M.Z., Folini, D., Wild, M., and Schär, C. (2012). Impact of Greenland's topographic height on precipitation and snow accumulation in idealized simulations. *Journal of Geophysical Research: Atmospheres*, *117*(D9), 7436.

Hall, R.E. (1988). Intertemporal substitution in consumption. *Journal of Political Economy*, *96*(2), 339-357.

Hanasaki, N., Fujimori, S., Yamamoto, T., Yoshikawa, S., Masaki, Y., Hijioka, Y., Kainuma, M., Kanamori, Y., Masui, T., Takahashi, K., and Kanae, S. (2013). A global water scarcity assessment under shared socio-economic pathways—Part 2: Water availability and scarcity. *Hydrological Earth System Science*, *17*, 2393-2413.

Hansen, J., Johnson, D., Lacis, A., Lebedeff, S., Lee, P., Rind, D., and Russell, G. (1981). Climate impact of increasing atmospheric carbon dioxide. *Science*, *213*(4511), 957-966.

Hansen, J., Lacis, A., Rind, D., Russell, G., Stone, P., Fung, I., Ruedy, R., and Lerner, J. (1984). Climate sensitivity: Analysis of feedback mechanisms. *Geophysical Monograph*, *29*(5), 130-163.

Haraden, J. (1992). An improved shadow price for CO_2. *Energy*, *17*(5), 419-426.

Harrison, P.A., Dunford, R.W., Holman, I.P., and Rounsevell, M.D.A. (2016). Climate change impact modelling needs to include cross-sectoral interactions. *Nature Climate Change*, *6*, 885-890.

Harvey, C.R. (1988). The real term structure and consumption growth. *Journal of Financial Economics*, *22*(2), 305-333.

Hauri, C., Gruber, N., Vogt, M., Doney, S.C., Feely, R.A., Lachkar, Z., Leinweber, A., McDonnell, A.M.P., Munnich, M., and Plattner, G.-K. (2013). Spatiotemporal variability and long-term trends of ocean acidification in the California Current System. *Biogeosciences*, *10*, 193-216.

Heal, G., and Millner, A. (2014). Uncertainty and decision making in climate change economics. *Review of Environmental Economics and Policy*, *8*(1), 120-137.

Hejazi, M., Edmonds, J., Clarke, L., Kyle, P., Davies, E., Chaturvedi, V., Wise, M., Patel, P., Eom, J., and Calvin, K. (2014). Integrated assessment of water scarcity over the 21st century under multiple climate change mitigation policies. *Hydrology and Earth System Sciences*, *18*, 2859-2883.

Held, I.M., Winton, M., Takahashi, K., Delworth, T., Zeng, F.R., and Vallis, G.K. (2010). Probing the fast and slow components of global warming by returning abruptly to preindustrial forcing. *Journal of Climate*, *23*, 2418-2427.

Herrington, T., and Zickfeld, K. (2014). Path independence of climate and carbon cycle response over a broad range of cumulative carbon emissions. *Earth System Dynamics*, *5*, 409-422.

Hirota, M., Holmgren, M., Van Nes, E.H., and Scheffer, M. (2011). Global resilience of tropical forest and savanna to critical transitions. *Science*, *334*(6053), 232-235.

Hitz, S., Smith, J., (2004). Estimating global impacts from climate change. *Global Environmental Change*, *14*(2004) 201-218.

Hodgson, D., and Miller, K. (1995). Modelling U.K. energy demand. In T. Barker, P. Ekins, and N. Johnstone (Eds.), *Global Warming and Energy Demand* (pp. 172-187). London, U.K.: Routledge.

Hoozemans, F.M.J., Marchand, M., and Pennekamp, H.A. (1993). *A Global Vulnerability Analysis: Vulnerability Assessment for Population, Coastal Wetlands and Rice Production and a Global Scale* (2nd Edition). Delft, The Netherlands: Delft Hydraulics.

Houser, T., Hsiang, S., Kopp, R., Larsen, K., Delgado, M., Jina, A., Mastrandrea, M., Mohan, S., Muir-Wood, R., Rasmussen, D.J., and Wilson, P. (2015*). Economic Risks of Climate Change: An American Prospectus*. New York: Columbia University Press.

Howard, P.H. (2014). *Omitted Damages: What's Missing from the Social Cost of Carbon*. Available: http://ww.policyintegrity.org/files/publications/Omitted_Damages_Whats_Missing_From_the_Social_Cost_of_Carbon.pdf [January 2017].

Howard, P.H., and Schwartz, J. (2016). *Think Global: International Reciprocity as Justification for a Global Social Cost of Carbon*. Available: https://ssrn.com/abstract=2822513 [February 2017].

Howard, P.H., and Sylvan, D. (2016). *The Wisdom of the Economic Crowd: Calibrating Integrated Assessment Models Using Consensus*. Available: http://ageconsearch.umn.edu/bitstream/235639/2/HowardSylvan_AAEA2016.pdf [October 2016].

Hsiang, S.M., and Narita, D. (2012). Adaptation to cyclone risk: Evidence from the global cross-section. *Climate Change Economics*, 3(2), 1-28.

Hsiang, S.M., M. Burke, and E. Miguel (2013), Quantifying the influence of climate on human conflict. *Science, 341*(6151), 1235367.

Hsiang, S.M., and Jina, A.J. (2014). *The Causal Effect of Environmental Catastrophe on Long-Run Economic Growth: Evidence from 6,700 Cyclones*. Technical report, National Bureau of Economic Research. Available: http://www.nber.org/papers/w20352 [April 2017].

Huybers, P. (2010). Compensation between model feedbacks and curtailment of climate sensitivity. *Journal of Climate, 23*, 3009-3018.

Intergovernmental Panel on Climate Change. (2007a). *Climate Change 2007: Impacts, Adaptation and Vulnerability. Working Group II Contribution to the Fourth Assessment Report of the Intergovernmental Panel on Climate Change*. M.L. Parry, O.F. Canziani, J.P. Palutikof, P.J. van der Linden, and C.E. Hanson (Eds.). Cambridge, U.K. and New York: Cambridge University Press.

Intergovernmental Panel on Climate Change. (2007b). *Climate Change 2007: Mitigation of Climate Change. Working Group III Contribution to the Fourth Assessment Report of the Intergovernmental Panel on Climate Change*. B. Metz, O.R. Davidson, P.R. Bosch, R. Dave, and L.A. Meyer (Eds.). Cambridge, U.K. and New York: Cambridge University Press.

Intergovernmental Panel on Climate Change. (2013). *Climate Change 2013: The Physical Science Basis. Summary for Policymakers. Contribution of Working Group I to the Fifth Assessment Report of the Intergovernmental Panel on Climate Change*. T.F. Stocker, , D. Qin, G.-K. Plattner, M. Tignor, S.K. Allen, J. Boschung, A. Nauels, Y. Xia, V. Bex and P.M. Midgley (Eds.). Cambridge, U.K. and New York: Cambridge University Press. Available: https://www.ipcc.ch/pdf/assessment-report/ar5/wg1/WGIAR5_SPM_brochure_en.pdf [October 2016].

Intergovernmental Panel on Climate Change. (2014a). *Climate Change 2014: Impacts, Adaptation, and Vulnerability. Working Group II Contribution to the Fifth Assessment Report of the Intergovernmental Panel on Climate Change*. C.B. Field, V.R. Barros, and L.L. White (Eds.). Cambridge, U.K. and New York: Cambridge University Press.

Intergovernmental Panel on Climate Change. (2014b). *Climate Change 2014: Mitigation of Climate Change. Working Group II: Impacts, Adaptation and Vulnerability*. Cambridge, U.K. and New York: Cambridge University Press.

Intergovernmental Panel on Climate Change. (2014c). *Climate Change 2014: Mitigation of Climate Change. Working Group III: Mitigation of Climate Change.* Cambridge, U.K. and New York: Cambridge University Press.

Interagency Working Group on the Social Cost of Carbon. (2010). *Technical Support Document: Social Cost of Carbon for Regulatory Impact Analysis Under Executive Order 12866* (February 2010). Available: https://obamawhitehouse.archives.gov/sites/default/files/omb/inforeg/for-agencies/Social-Cost-of-Carbon-for-RIA.pdf [January 2017].

Interagency Working Group on the Social Cost of Carbon. (2013a). *Technical Support Document: Technical Update of the Social Cost of Carbon for Regulatory Impact Analysis Under Executive Order 12866* (May 2013). Available: https://obamawhitehouse.archives.gov/sites/default/files/omb/inforeg/social_cost_of_carbon_for_ria_2013_update.pdf [January 2017].

Interagency Working Group on the Social Cost of Carbon. (2013b). *Technical Support Document: Technical Update of the Social Cost of Carbon for Regulatory Impact Analysis Under Executive Order 12866* (November 2013 Revision). Available: https://obamawhitehouse.archives.gov/sites/default/files/omb/assets/inforeg/technical-update-social-cost-of-carbon-for-regulator-impact-analysis.pdf [January 2017].

Interagency Working Group on the Social Cost of Carbon. (2015a). *Technical Support Document: Technical Update of the Social Cost of Carbon for Regulatory Impact Analysis Under Executive Order 12866* (July 2015 Revision). Available: https://obamawhitehouse.archives.gov/sites/default/files/omb/inforeg/scc-tsd-final-july-2015.pdf [January 2017].

Interagency Working Group on the Social Cost of Carbon. (2015b). *Response to Comments: Technical Update of the Social Cost of Carbon for Regulatory Impact Analysis Under Executive Order 12866* (July 2015). Available: https://obamawhitehouse.archives.gov/sites/default/files/omb/inforeg/scc-response-to-comments-final-july-2015.pdf [January 27, 2017].

Interagency Working Group on the Social Cost of Greenhouse Gases. (2016a). *Addendum to Technical Support Document on Social Cost of Carbon for Regulatory Impact Analysis under Executive Order 12866: Application of the Methodology to Estimate the Social Cost of Methane and the Social Cost of Nitrous Oxide.* Available: https://obamawhitehouse.archives.gov/sites/default/files/omb/inforeg/august_2016_sc_ch4_sc_n2o_addendum_final_8_26_16.pdf [January 2017].

Interagency Working Group on the Social Cost of Greenhouse Gases. (2016b). *Technical Support Document: Technical Update of the Social Cost of Carbon for Regulatory Impact Analysis Under Executive Order 12866* (September 2016 Revision). Washington, DC: Interagency Working Group on the Social Cost of Carbon. Available: https://obamawhitehouse.archives.gov/sites/default/files/omb/inforeg/august_2016_sc_ch4_sc_n2o_addendum_final_8_26_16.pdf [January 2017].

Isaac, M., and van Vuuren, D.P. (2009). Modeling global residential sector energy demand for heating and air conditioning in the context of climate change. *Energy Policy, 37*(2), 507-521.

Jevrejeva, S., Grinsted, A., and Moore, J.C. (2014). Upper limit for sea level projections by 2100. *Environmental Research Letters, 9*(10), 104008.

Joint Global Change Research Institute. (2015). *Climate Change Assessment Model* (v. 4.2). Richland, WA: Pacific Northwest National Laboratory.

Jones, C., Robertson, E., Arora, V., and others (2013). 21st Century compatible CO2 emissions and airborne fraction simulated by CMIP5 Earth System models under four representative concentration pathways. *Journal of Climate, 26,* 4398-4413.

Joos, F., Roth, R., Fuglestvedt, J.S., Peters, G.P., Enting, I.G., Bloh, W. von, Brovkin, V., Burke, E.J., Eby, M., Edwards, N.R., Friedrich, T., Frölicher, T.L., Halloran, P.R., Holden, P.B., Jones, C., Kleinen, T., Mackenzie, F.T., Matsumoto, K., Meinshausen, M., Plattner, G.K., Reisinger, A., Segschneider, J., Shaffer, G., Steinacher, M., Strassmann, K., Tanaka, K., Timmermann, A., and Weaver, A.J. (2013). Carbon dioxide and climate impulse response functions for the computation of greenhouse gas metrics: A multi-model analysis. *Atmospheric Chemistry and Physics, 13*, 2793-2825.

Kane, S., Reilly, J.M., and Tobey, J. (1992). An empirical study of the economic effects of climate change on world agriculture. *Climatic Change, 21*(1), 17-35.

Kay, J.E., Deser, C., Phillips, A., Mai, A., Hannay, C., Strand, G., Arblaster, J.M., Bates, S.C., Danabasoglu, G., Edwards, J., Holland, M., Kushner, P., Lamarque, J.-F., Lawrence, D., Lindsay, K., Middleton, A., Munoz, E., Neale, R., Oleson, K., Polvani, L., and Vertenstein, M. (2014). The Community Earth System Model (CESM) Large Ensemble Project: A community resource for studying climate change in the presence of internal climate variability. *Bulletin of the American Meteorological Society, 96*, 1333-1349.

Keller, K., Bolker, B.M., and Bradford, D.F. (2004). Uncertain climate thresholds and optimal economic growth. *Journal of Environmental Economics and Management, 48*(1), 723-741.

Kim, S.H., Hejazi, M., Liu, L., Calvin, K., Clarke, L., Edmonds, J., Kyle, P., Patel, P., Wise, M., and Davies, E. (2016). Balancing global water availability and use at basin scale in an integrated assessment model. *Climatic Change, 136*(2), 217-231.

Kitous, K. (2006). *Web Documentation by Enerdata Services of the POLES Model.* Available: http://www.eie.gov.tr/projeler/document/5_POLES_description.pdf [October 2016].

Klein, A.-M., Vaissière, B.E., Cane, J.H., Steffan-Dewenter, I., Cunningham, S.A., Kremen, C., and Tscharntke, T. (2007). Importance of pollinators in changing landscapes for world crops. *Proceedings of the Royal Society of London B: Biological Sciences, 274*(1608), 303-313.

Knutti, R., and Rugenstein, M.A.A. (2015). Feedbacks, climate sensitivity and the limits of linear models. *Philosophical Transactions of the Royal Society A: Mathematical, Physical and Engineering Sciences, 373*(2054).

Kolstad, C., Urama, K., Broome, J., Bruvoll, A., Cariño-Olvera, M., Fullerton, D., Gollier, C., Hanemann, W.M., Hassan, R., Jotzo, F., Khan, M.R., Meyer, L., and Mundaca, L. (2014). Social, economic and ethical concepts and methods. In O. Edenhofer, R. Pichs-Madruga, Y. Sokona, E. Farahani, S. Kadner, K. Seyboth, A. Adler, I. Baum, S. Brunner, P. Eickemeier, B. Kriemann, J. Savolainen, S. Schlömer, C. von Stechow, T. Zwickel, and J.C. Minx (Eds.), *Climate Change 2014: Mitigation of Climate Change. Working Group III Contribution to the Fifth Assessment Report of the Intergovernmental Panel on Climate Change* (Ch. 3) (pp. 173-248). Cambridge, U.K. and New York: Cambridge University Press.

Kopp, R.E., and Mignone, B.K. (2012). The U.S. government's social cost of carbon estimates after their first two years: Pathways for improvement. *Economics, 6*(2012-15), 1-41.

Kopp, R.E., Golub, A., Keohane, N.O., and Onda, C. (2012). The influence of the specification of climate change damages on the social cost of carbon. *Economics, 6*(2012-13), 1-40.

Kopp, R.E., and Mignone, B.K. (2013). Circumspection, reciprocity, and optimal carbon prices. *Climatic Change 120*(4), 831-843.

Kopp, R.E., Horton, R.M., Little, C.M., Mitrovica, J.X., Oppenheimer, M., Rasmussen, D.J., Strauss, B.H., and Tebaldi, C. (2014). Probabilistic 21st and 22nd century sea-level projections at a global network of tide gauge sites. *Earth's Future, 2*(8), 383-406.

Kopp, R.E., Hay, C.C., Little, C.M., and Mitrovica, J.X. (2015). Geographic variability of sea-level change. *Current Climate Change Reports, 1*(3), 192-204.

Kopp, R.E., Kemp, A.C., Bittermann, K., Horton, B.P., Donnelly, J.P., Gehrels, W.R., Hay, C.C., Mitrovica, J.X., Morrow, E.D., and Rahmstorf, S. (2016a). Temperature-driven global sea-level variability in the Common Era. *Proceedings of the National Academy of Sciences of the United States of America, 113*(1), E1434-E1441.

Kopp, R.E., Shwom, R., Wagner, G., and Yuan, J. (2016b). *Tipping Elements, Tipping Points, and Economic Catastrophes: Implications for the Cost of Climate Change.* New Brunswick, NJ: Rutgers University.

Kraucunas I., Clarke, L., Dirks, J. Hathaway, J., Hejazi, M., Hibbard, K., Huang, M., Jin, C., Kintner-Meyer, M., Kleese van Dam, K., Leung, R., Li, H.-Y., Moss, R., Peterson, M., Rice, J., Scott, M., Thomson, A., Voisin, N., and West, T. (2015). Investigating the nexus of climate, energy, water, and land at decision-relevant scales: The platform for regional integrated modeling and analysis (PRIMA). *Climatic Change, 129*(3-4), 573-588.

Kriegler, E., Edmonds, J., Hallegatte, S., Ebi, K.L., Kram, T., Riahi, K., Winkler, H., and van Vuuren, D.P. (2014). A new scenario framework for climate change research: The concept of shared climate policy assumptions. *Climatic Change, 122*(3), 401-414.

Kunreuther H., S. Gupta, V. Bosetti, R. Cooke, V. Dutt, M. Ha-Duong, H. Held, J. Llanes-Regueiro, A. Patt, E. Shittu, and E. Weber. (2014). Integrated Risk and Uncertainty Assessment of Climate Change Response Policies. In O. Edenhofer, R. Pichs-Madruga, Y. Sokona, E. Farahani, S. Kadner, K. Seyboth, A. Adler, I. Baum, S. Brunner, P. Eickemeier, B. Kriemann, J. Savolainen, S. Schlömer, C. von Stechow, T. Zwickel and J.C. Minx (Eds.), *Climate Change 2014: Mitigation of Climate Change. Contribution of Working Group III to the Fifth Assessment Report of the Intergovernmental Panel on Climate Change.* Cambridge, U.K. and New York: Cambridge University Press.

Kyle, P., Mueller, C., Calvin, K., and Thomson, A.M. (2014). Meeting the radiative forcing targets of the representative concentration pathways in a world with agricultural climate impacts. *Earth's Future, 2*(2), 83-98.

Labriet, M., Joshi, S., Babonneau, F., Edwards, N., Holden, P., Kanudia, A., Loulou, R., and Vielle, M. (2013). Worldwide impacts of climate change on energy for heating and cooling. *Mitigation and Adaptation Strategies for Global Change, 20*(7), 1111-1136.

Leatherman, S.P., and Nicholls, R.J. (1995). Accelerated sea-level rise and developing countries: An overview. *Journal of Coastal Research, 14*, 1-14.

Leggett, J., W.J. Pepper and R.J. Swart. (1992). Emissions scenarios for the IPCC: An update. In J.T. Houghton, B.A. Callander, and S.K. Varney (Eds.). *Climate Change 1992: The Supplementary Report to the IPCC Scientific Assessment* (pp. 75-95). Cambridge, U.K.: Cambridge University Press.

Lemoine, D., and Traeger, C.P. (2016). Economics of tipping the climate dominoes. *Nature Climate Change, 6*, 514-519.

Lenton, T.M., Held, H., Kriegler, E., Hall, J.W., Lucht, W., Rahmstorf, S., and Schellnhuber, H.J. (2008). Tipping elements in the Earth's climate system. *Proceedings of the National Academy of Sciences of the United States of America, 105*(6), 1786-1793.

Lewis, N., and Curry, J.A. (2014). The implications for climate sensitivity of AR5 forcing and heat uptake estimates. *Climate Dynamics, 45*(3), 1009-1023.

Li, J., Mullan, M., and Helgeson. J. (2014). Improving the practice of economic analysis of climate change adaptation. *Journal of Benefit-Cost Analysis, 5*(3), 445-467.

Libardoni, A.G., and Forest, C.E. (2011). Sensitivity of distributions of climate system properties to the surface temperature dataset. *Geophysical Research Letters, 38*(22).

Libardoni, A.G., and Forest, C.E. (2013). Correction to "Sensitivity of distributions of climate system properties to the surface temperature data set." *Geophysical Research Letters, 40*(10), 2309-2311.

Lind, R. (1982). *Discounting for Time and Risk in Energy Policy.* Baltimore, MD: Johns Hopkins University Press for Resources for the Future.

Link, P.M., and Tol, R.S.J. (2004). Possible economic impacts of a shutdown of the thermohaline circulation: An application of FUND. *Portuguese Economic Journal, 3*, 99-114.

LoGiudice, K., Ostfeld, R.S., Schmidt, K.A., and Keesing, F. (2003). The ecology of infectious disease: Effects of host diversity and community composition on Lyme disease risk. *Proceedings of the National Academy of Sciences of the United States of America, 100*(2), 567-571.

Lutz, W., Butz, W., and S. KC (Eds.). (2014). *World Population and Human Capital in the Twenty-First Century.* Oxford, U.K.: Oxford University Press.

Manabe, S., and Stouffer, R.J. (1980). Sensitivity of a global climate model to an increase of CO_2 concentration in the atmosphere. *Journal of Geophysical Research: Oceans, 85*(C10), 5529-5554.

Manabe, S., and Wetherald, R.T. (1967). Thermal equilibrium of the atmosphere with a given distribution of relative humidity. *Journal of the Atmospheric Sciences, 24*(3), 241-259.

Manabe, S., and Wetherald, R.T. (1975). The effects of doubling the CO_2 concentration on the climate of a general circulation model. *Journal of Atmospheric Sciences, 32*(1), 3-15.

Manabe, S., and Wetherald, R.T. (1980). On the distribution of climate change resulting from an increase in CO_2 content of the atmosphere. *Journal of the Atmospheric Sciences, 37,* 99-118.

Mankiw, G. (1981). The permanent income hypothesis and the real interest rate. *Economics Letters, 7*(4), 307-311.

Mann, M.E., Zhang, Z., Rutherford, S., Bradley, R.S., Hughes, M.K., Shindell, D., Ammann, C., Faluvegi, G., and Ni, F. (2009). Global signatures and dynamical origins of the little ice age and medieval climate anomaly. *Science, 326*(5957), 1256-1260.

Manne, A.S., and Richels, R.G. (1994). The costs of stabilizing global CO_2 emissions: A probabilistic analysis based on expert judgment. *Energy Journal, 15*(1), 31-56.

Marcott, S.A., Shakun, J.D., Clark, P.U. and Mix, A.C. (2013). A reconstruction of regional and global temperature for the past 11,300 years. *Science, 339,* 1198-1201.

Marten, A.L., Kopp, R.E., Shouse, K.C., Griffiths, C.W., Hodson, E.L., Kopits, E., Mignone, B.K., Moore, C., Newbold, S.C., Waldhoff, S., and Wolverton, A. (2013). Improving the assessment and valuation of climate change impacts for policy and regulatory analysis. *Climatic Change, 117*(3), 433-438.

Martens, W.J.M. (1998). Climate change, thermal stress and mortality changes. *Social Science and Medicine, 46*(3), 331-344.

Martens, W.J.M., Jetten, T.H., Rotmans, J., and Niessen, L.W. (1995). Climate change and vector-borne diseases: A global modelling perspective. *Global Environmental Change, 5*(3), 195-209.

Martens, W.J.M., Jetten, T.H., and Focks, D.A. (1997). Sensitivity of malaria, schistosomiasis and dengue to global warming. *Climatic Change, 35*(2), 145-156.

Martin, P.H., and Lefebvre, M.G. (1995). Malaria and climate: Sensitivity of malaria potential transmission to climate. *AMBIO: A Journal of the Human Environment, 24*(4), 200-207.

Mastrandrea, M.D., Field, C.B., Stocker, T.F., Edenhofer, O., Ebi, K.L., Frame, D.J., Held, H., Kriegler, E., Mach, K.J., and Matschoss, P.R. (2010). *Guidance Note for Lead Authors of the IPCC Fifth Assessment Report on Consistent Treatment of Uncertainties.* Geneva, Switzerland: Intergovernmental Panel on Climate Change. Available: http://www.ipcc.ch/pdf/supporting-material/uncertainty-guidance-note.pdf [October 2016].

Mathis, J.T., Cooley, S.R., Lucey, N., Colt, S., Ekstrom, J., Hurst, T., Hauri, C., Evans, W., Cross, J.N., and Feely, R.A. (2015). Ocean acidification risk assessment for Alaska's fishery sector. *Progress in Oceanography: Synthesis of Arctic Research (SOAR), 136,* 71-91.

Matsuoka Y., Kainuma, M., and Morita, T. (1995). Scenario analysis of global warming using the Asian Pacific Integrated Model (AIM). *Energy Policy, 23*(4/5), 357-371.

McNeil, B.I., and Matear, R.J. (2008). Southern Ocean acidification: A tipping point at 450 ppm atmospheric CO_2. *Proceedings of the National Academy of Sciences of the United States of America, 105*(48), 18860-18864.

Meehl, G.A., Stocker, T.F., Collins, W.D., Friedlingstein, P., Gaye, A.T., Gregory, J.M., Kitch, A., Knutti, R., Murphy, J.M., Noda, A., Raper, S.C.B., Watterson, I.G., Weaver, A.J., and Zhao, Z.-C. (2007). Global climate projections. In S. Solomon, D. Qin, M. Manning, Z. Chen, M. Marquis, K.B. Averyt, M. Tignor, and H.L. Miller (Eds.), *Climate Change 2007: The Physical Science Basis* (pp. 747-845). Cambridge, U.K. and New York: Cambridge University Press.

Meinshausen, M., Meinshausen, N., Hare, W., Raper, S.C., Frieler, K., Knutti, R., Frame, D.J., and Allen, M.R. (2009). Greenhouse-gas emission targets for limiting global warming to 2 °C. *Nature, 458,* 1158-1162.

Mengel, M., Levermann, A., Frieler, K., Robinson, A., Marzeion, B., and Winkelmann, R. (2016). Future sea level rise constrained by observations and long-term commitment. *Proceedings of the National Academy of Sciences of the United States of America, 113*(10), 2597-2602.

Meraner, K., Mauritsen, T., and Voigt, A. (2013). Robust increase in equilibrium climate sensitivity under global warming. *Geophysical Research Letters, 40*(22), 5944-5948.

Millar, R.J., Otto, A., Forster, P.M., Lowe, J.A., Ingram, W.J., and Allen, M.R. (2015). Model structure in observational constraints on transient climate response. *Climatic Change, 131*(2), 199-211.

Millar, R.J., Nicholls, Z.R., Friedlingstein, P., and Allen, M.R. (2016). A modified impulse-response representation of the global response to carbon dioxide emissions. *Atmospheric Chemistry and Physics,* 1-20.

Milne, G.A., Gehrels, W.R., Hughes, C.W., and Tamisiea, M.E. (2009). Identifying the causes of sea-level change. *Nature Geoscience, 2,* 471-478.

Mima, S., and Criqui, P. (2009). *Assessment of the Impacts Under Future Climate Change on the Energy Systems with the POLES model.* Paper Presented at 2009 International Energy Workshop, June 17-19, Venice, Italy.

Mitchell, T.D. (2003). Pattern scaling: An examination of the accuracy of the technique for describing future climates. *Climatic Change, 60*(3), 217-242.

Moore, F., Lantz Baldos, U., Hertel, T., and Diaz, D. (2016). *Welfare Changes from Climate Change Impacts on the Agricultural Sector: New Damage Functions from Over 1000 Yield Studies.* Presented at the 19th Annual Conference on Global Economic Analysis, Washington, DC. Available: https://www.gtap.agecon.purdue.edu/resources/res_display.asp?RecordID=5056 [December 2016].

Morita, T., Kainuma, M., Harasawa, H., Kai, K., Dong-Kun, L., and Matsuoka, Y. (1994). *Asian-Pacific Integrated Model for Evaluating Policy Options to Reduce Greenhouse Gas Emissions and Global Warming Impacts.* Tsukuba, Japan: National Institute for Environmental Studies.

Moss, R.H., Edmonds, J.A., Hibbard, K.A., Manning, M.R., Rose, S.K., van Vuuren, D.P., Carter, T.R., Emori, S., Kainuma, M., Kram, T., Meehl, G.A., Mitchell, J.F.B., Nakicenovic, N., Riahi, K., Smith, S.J., Stouffer, R.J., Thomson, A.M., Weyant, J.P., and Wilbanks, T.W. (2010). The next generation of scenarios for climate change research and assessment. *Nature, 463,* 747-756.

Mueller, U.K., and Watson, M.W. (2016). Measuring uncertainty about long-run predictions. *Review of Economic Studies, 83*(4), 1711-1740.

Murphy, K.M., and Topel, R.H. (2013). Some basic economics of national security. *American Economic Review, 103*(3), 508-511.

Murray, C.J.L., and Lopez, A.D. (Eds.). (1996). *The Global Burden of Disease.* Cambridge, MA: Harvard School of Public Health, Harvard University Press.

Myhre, G., Shindell, D., Bréon, F.M., Collins, W., Fuglestvedt, J., Huang, J., Koch, D., Lamarque, J.F., Lee, D., Mendoza, B., Nakajima, T., Robock, A., Stephens, G., Takemura, T., and Zhang, H. (2013). Anthropogenic and natural radiative forcing. In T.F. Stocker, D. Qin, G.-K. Plattner, M. Tignor, S.K. Allen, J. Boschung, A. Nauels, Y. Xia, V. Bex, and P.M. Midgley (Eds.), *Climate Change 2013: The Physical Science Basis. Working Group I Contribution to the Fifth Assessment Report of the Intergovernmental Panel on Climate Change* (pp. 659-740). Cambridge, U.K. and New York: Cambridge University Press.

Nakicenovic, N., Alcamo, J., Davis, G., de Vries, B., Fenhann, J., Gaffin, S., Gregory, K., Grübler, A., Jung, T.Y., Kram, T., La Rovere, E. L., Michaelis, L., Mori, S., Morita, T., Pepper, W., Pitcher, H., Price, L., Riahi, K., Roehrl, A., Rogner, H.-H., Sankovski, A., Schlesinger, M., Shukla, P., Smith, S., Swart, R., van Rooijen, S., Victor, N., and Dadi, Z. (2000). *Special Report on Emissions Scenarios: A Special Report of Working Group III of the Intergovernmental Panel on Climate Change.* Cambridge, U.K.: Cambridge University Press. Available: http://www.grida.no/climate/ipcc/emission/index.htm [October 2016].

Nam, K., Waugh, C.J., Paltsev, S., Reilly, J.M., Karplus, V.J., (2014). Synergy between pollution and carbon emissions control: comparing China and the United States. *Energy Economics, 46,* 186-201.

Narita, D., Rehdanz, K., and Tol, R.S.J. (2012). Economic costs of ocean acidification: A look into the impacts on global shellfish production. *Climatic Change, 113*(3), 1049-1063.

National Academy of Sciences. (1979). *Carbon Dioxide and Climate: A Scientific Assessment.* Washington, DC: National Academy Press.

National Research Council. (2010). *Hidden Costs of Energy: Unpriced Consequences of Energy Production and Use.* Committee on Health, Environmental, and Other External Costs and Benefits of Energy Production and Consumption; Board on Environmental Studies and Toxicology; Division on Earth and Life Studies; Board on Energy and Environmental Systems; Division on Engineering and Physical Sciences; Board on Science, Technology, and Economic Policy; Policy and Global Affairs. Washington, DC: The National Academies Press.

National Research Council. (2012). *A National Strategy for Advancing Climate Modeling.* Committee on a National Strategy for Advancing Climate Modeling; Division on Earth and Life Studies; Board on Atmospheric Sciences and Climate. Washington, DC: The National Academies Press.

National Research Council. (2013). *Abrupt Impacts of Climate Change: Anticipating Surprises.* Committee on Understanding and Monitoring Abrupt Climate Change and its Impacts, Board on Atmospheric Sciences and Climate Division on Earth and Life Studies. Washington, DC: The National Academies Press. Available: https://www.nap.edu/catalog/18373/abrupt-impacts-of-climate-change-anticipating-surprises [January 2017].

National Academies of Sciences, Engineering, and Medicine. (2016). *Assessment of Approaches to Updating the Social Cost of Carbon: Phase 1 Report on a Near-Term Update.* Committee on Assessing Approaches to Updating the Social Cost of Carbon, Board on Environmental Change and Society. Washington, DC: The National Academies Press.

Nelson, G.C., Valin, H., Sands, R.D., Havlík, P., Ahammad, H., Deryng, D., Elliott, J.F., Fujimori, S., Hasegawa, T., Heyhoe, E., Kyle, P., Von Lampe, M., Lotze-Campen, H., Mason d'Croz, D., van Meijl, H., van der Mensbrugghe, D., Müller, C., Popp, A., Robertson, R., Robinson, S., Schmid, E., Schmitz, C., Tabeau, A., and Willenbockel, D. (2014). Climate change effects on agriculture: Economic responses to biophysical shocks. *Proceedings of the National Academy of Sciences of the United States of America, 111*(9), 3274-3279.

Newell, R., and Pizer, W. (2003). Discounting the distant future: How much do uncertain rates increase valuations? *Journal of Environmental Economics and Management, 46*(1), 52-71.

Nicholls, R.J., and Leatherman, S.P. (1995). The implications of accelerated sea-level rise for developing countries: A discussion. *Journal of Coastal Research, 14*, 303-323.

Nicholls, R.J., Tol, R.S.J., and Vafeidis, A.T. (2008). Global estimates of the impact of a collapse of the west Antarctic ice sheet: An application of FUND. *Climate Change, 91*(1-2), 171-191.

Nordhaus, W. (1982). How fast should we graze the global commons? *American Economic Review, 72*(2), 242-246.

Nordhaus, W.D. (1991). To slow or not to slow: The economics of the greenhouse effect. *Economic Journal, 101*(407), 920-937.

Nordhaus, W.D. (1994a). Expert opinion on climatic change. *American Scientist, 82*, 45-51.

Nordhaus, W.D. (1994b). *Managing the Global Commons: The Economics of Climate Change.* Cambridge, MA: MIT Press.

Nordhaus, W.D. (2007). *Accompanying Notes and Documentation on Development of DICE-2007 Model: Notes on DICE-2007.delta.v8 as of September 21, 2007.* New Haven, CT: Yale University Press. Available: http://www.econ.yale.edu/~nordhaus/homepage/Accom_Notes_100507.pdf [October 2016].

Nordhaus, W.D. (2008). *A Question of Balance: Weighing the Options on Global Warming Policies.* New Haven, CT: Yale University Press.

Nordhaus, W.D. (2010). Economic aspects of global warming in a post-Copenhagen environment. *Proceedings of the National Academy of Sciences of the United States of America, 107*(26), 11721-11726.

Nordhaus, W.D. (2011). *Estimates of the Social Cost of Carbon: Background and Results from the RICE-2011 Model.* Working Paper 17540. Cambridge, MA: National Bureau of Economic Research.

Nordhaus, W.D. (2014). Estimates of the social cost of carbon: Concepts and results from the DICE-2013R model and alternative assumptions. *Journal of the Association of Environmental and Resource Economists, 1*(1/2), 273-312.

Nordhaus, W.D., and Boyer, J. (2000). *Warming the World: Economic Models of Global Warming.* Cambridge, MA: MIT Press.

Nordhaus, W.D., and Popp, D. (1997). What is the value of scientific knowledge? An application to global warming using the PRICE model. *The Energy Journal*, 1-45.

O'Neill, B.C., Kriegler, E., Riahi, K., Ebi, K.L., Hallegatte, S., Carter, T.R., Mathur, R., and van Vuuren, D.P. (2014). A new scenario framework for climate change research: The concept of shared socioeconomic pathways. *Climatic Change, 122*(3), 387-400.

Oppenheimer, M., Campos, M., Warren, R., Birkmann, J., Luber, G., O'Neill, B., and Takahashi, K. (2014). Emergent risks and key vulnerabilities. In C.B. Field, V.R. Barros, D.J. Dokken, K.J. Mach, M.D. Mastrandrea, T.E. Bilir, M. Chatterjee, K.L. Ebi, Y.O. Estrada, R.C. Genova, B. Girma, E.S. Kissel, A.N. Levy, S. MacCracken, P.R. Mastrandrea, and L.L. White (Eds.), *Climate Change 2014: Impacts, Adaptation, and Vulnerability. Part A: Global and Sectoral Aspects. Contribution of Working Group II to the Fifth Assessment Report of the Intergovernmental Panel on Climate Change* (pp. 1039-1099). Cambridge, U.K. and New York: Cambridge University Press. Available: http://www.ipcc.ch/pdf/assessment-report/ar5/wg2/WGIIAR5-Chap19_FINAL.pdf [January 2017].

Orr, J.C., Fabry, V.J., Aumont, O., Bopp, L., Doney, S.C., Feely, R.A., Gnanadesikan, A., Gruber, N., Ishida, A., Joos, F., Key, R.M., Lindsay, K., Maier-Reimer, E., Matear, R., Monfray, P., Mouchet, A., Najjar, R.G., Plattner, G.K., Rodgers, K.B., Sabine, C.L., Sarmiento, J.L., Schlitzer, R., Slater, R.D., Totterdell, I.J., Weirig, M.F., Yamanaka, Y., and Yool, A. (2005). Anthropogenic ocean acidification over the twenty-first century and its impact on calcifying organisms, *Nature, 437*(7059), 681-686.

Otto, A., Todd, B.J., Bowerman, N., Frame, D.J., and Allen, M.R. (2013). Climate system properties determining the social cost of carbon. *Environmental Research Letters, 8*(2), 024032.

PALAEOSENS Project. (2012). Making sense of palaeoclimate sensitivity. *Nature, 491,* 683-691.

Parry M., Arnell, N., Berry, P., Dodman, D., Fankhauser, S., Hope, C., Kovats, S., Nicholls, R., Satterthwaite, D., Tiffin, R., and Wheeler, T. (2009). *Assessing the Costs of Adaptation to Climate Change: A Review of the UNFCCC and Other Recent Estimates.* London, U.K.: International Institute for Environment and Development and Grantham Institute for Climate Change.

Pearce, D.W., and Moran, D. (1994). *The Economic Value of Biodiversity.* London, U.K.: EarthScan.

Peck, S.C., and Teisberg, T.J. (1993). Global warming uncertainties and the value of information: An analysis using CETA. *Resource and Energy Economics, 15*(1), 71-97.

Pepper. W.J., Leggett, J.A., Swart, R.J., Wasson, J., Edmonds, J., and Mintzer, I. (1992). *Emission Scenarios for the IPCC—An Update: Background Documentation on Assumptions, Methodology, and Results.* Washington, DC: U.S. Environmental Protection Agency.

Perez-Garcia, J., Joyce, L.A., Binkley, C.S., and McGuire, A.D. (1995). Economic impacts of climatic change on the global forest sector: An integrated ecological/economic assessment. *Journal Critical Reviews in Environmental Science and Technology, 27*(Suppl. 1), 123-138.

Pindyck, R.S. (2015). *The Use and Misuse of Models for Climate Policy.* Working Paper 21097. Cambridge, MA: National Bureau of Economic Research.

Pizer,W., M. Adler, J. Aldy, D. Anthoff, M. Cropper, K. Gillingham, M. Greenstone, B. Murray, R. Newell, R. Richels, A. Rowell, S. Waldhoff, J. Wiener. (2014). Using and improving the social cost of carbon. *Science, 346*(6214), 189-1190.

Pueyo, S. (2012). Solution to the paradox of climate sensitivity. *Climatic Change, 113*(2), 163-179.

Rahmstorf, S. (2007). A semi-empirical approach to projecting future sea-level rise. *Science, 315*(5810), 368-370.

Rasmussen, D.J., Meinshausen, M., and Kopp, R.E. (2016). Probability-weighted ensembles of U.S. county-level climate projections for climate risk analysis. *Journal of Applied Meteorology and Climatology, 55,* 2301-2322.

Reilly, J., Edmonds, J., Gardner, R., and Brenkert, A. (1987). Monte Carlo analysis of the IEA/ORAU Energy/Carbon Emissions Model. *Energy Journal, 8*(3), 1-29.

Reilly, J., and Richards, K. (1993). Climate change damage and the trace gas index issue. *Environmental & Resource Economics, 3*(1), 41-61.

Reilly, J.M., Hohmann, N., and Kane, S. (1994). Climate change and agricultural trade: Who benefits, who loses? *Global Environmental Change, 4*(1), 24-36.

Reilly, J., Paltsev, S., Felzer, B., Wang, X., Kicklighter, D., Melillo, J., Prinn, R., Sarofim, M., and Wang, C. (2007). Global economic effects of changes in crops, pasture, and forests due to changing climate, carbon dioxide, and ozone. *Energy Policy, 35*(11), 5370-5383.

Reilly, J., Melillo, J., Cai, Y., Kicklighter, D., Gurgel, A., Paltsev, S., Cronin, T., Sokolov, A., and Schlosser, A. (2012a). Using land to mitigate climate change: Hitting the target, recognizing the tradeoffs. *Environmental Science and Technology, 46*(11), 5672-5679.

Reilly, J., Paltsev, S., Strzepek, K., Selin, N., Cai, Y., Nam, H.-M., Monier, E., Dutkiewitz, S., Scott, J., Webster, M., and Sokolov, S. (2012b). *Valuing Climate Impacts in Integrated Assessment Models: The MIT IGSM.* Available: http://globalchange.mit.edu/files/document/MITJPSPGC_Rpt219.pdf [October 2016].

Revesz, R.L., Howard, P.H., Arrow, K., Goulder, L.H, Kopp, R.E., Livermore, M.A., Oppenheimer, M., and Sterner, T. (2014). Global warming: Improve economic models of climate change, *Nature, 508*(7495), 173-175.

Riahi, K., Edmonds, J., O'Neill, B., van Vuuren, D., Kriegler, E., Fujimori, J.S., Bauer, N., Calvin, K., Dellink, R., Fricko, O., Lutz, W., Popp, A., Cuaresma, J., Leimbach, M., Jiang, L., Kram, T., Rao, S., Emmerling, J., Ebi, K., Hasegawa, T., Havlik, P., Humpenöder, F., Da Silva, L., Smith, S., Stehfest, E., Bosetti, V., Eom, J., Gernaat, D., Masui, T., Rogelj, J., Strefler, J., Drouet, L., Krey, V., Luderer, G., Harmsen, M., Takahashi, K., Baumstark, L., Doelman, J., Kainuma, M., Klimont, Z., Marangoni, G., Lotze-Campen, H., Obersteiner, M., Tabeau, A., and Tavoni, M. (2016). The Shared Socioeconomic Pathways and their energy, land use, and greenhouse gas emissions implications: An overview. *Global Environmental Change*, 42, 153-168.

Richardson, M., Cowtan, K., Hawkins, E., and Stolpe, M.B. (2016). Reconciled climate response estimates from climate models and the energy budget of Earth. *Nature Climate Change*, 6, 931-935.

Ricke, K.L., and Caldeira, K. (2014). Maximum warming occurs about one decade after a carbon dioxide emission. *Environmental Research Letters*, 9(12), 124002.

Ringer, M.A., McAvaney, B.J., Andronova, N., Buja, L.E., Esch, M., Ingram, W.J., Li, B., Quaas, J., Roeckner, E., Senior, C.A., Soden, B.J., Volodin, E.M., Webb, M.J., and Williams, K.D. (2006). Global mean cloud feedbacks in idealized climate change experiments. *Geophysical Research Letters*, 33(7).

Rodolfo-Metalpa, R., Houlbrèque, F., Tambutté, É., Boisson, F., Baggini, C., Patti, F.P., Jeffree, R., Fine, M., Foggo, A., Gattuso, J.-P., and Hall-Spencer, J.M. (2011). Coral and mollusc resistance to ocean acidification adversely affected by warming. *Nature Climate Change*, 1, 308-312.

Roe, G.H., and Armour, K.C. (2011). How sensitive is climate sensitivity? *Geophysical Research Letters*, 38(14).

Roe, G.H., and Baker, M.B. (2007). Why is climate sensitivity so unpredictable? *Science*, 318(5850), 629-632.

Rose, S., Kriegler, E., Bibas, R., Calvin, K., Popp, A., van Vuuren, D.P., and Weyant, J. (2014a). Bioenergy in energy transformation and climate management. *Climatic Change*, 123(3-4), 477-493.

Rose, S., Turner, D., Blanford, G., Bistline, J., de la Chesnaye, F., and Wilson, T. (2014b). *Understanding the Social Cost of Carbon: A Technical Assessment*. Report 3002004657. Palo Alto, CA: Electric Power Research Institute.

Roson, R., and Sartori, M. (2010). *The ENVironmental Impact and Sustainability Applied General Equilibrium (ENVISAGE) Model—Introducing Climate Change Impacts and Adaptation*. Washington, DC: The World Bank.

Roson, R., and Sartori, M. (2016). *Estimation of Climate Change Damage Functions for 140 Regions in the GTAP9 Database*. Available: http://www.unive.it/media/allegato/DIP/Economia/Working_papers/Working_papers_2016/WP_DSE_roson_sartori_06_16.pdf [January 2017].

Roson, R., and van der Mensbrugghe, D. (2012). Climate change and economic growth: Impacts and interactions. *International Journal of Sustainable Economy*, 4(3), 270.

Rotmans, J. (1990). *IMAGE: An Integrated Model to Assess the Greenhouse Effect*. Dordrecht, The Netherlands: Kluwer Academic Publishers.

Sanderson, B.M., Knutti, R., and Caldwell, P. (2015). Addressing interdependency in a multi-model ensemble by interpolation of model properties. *Journal of Climate*, 28, 5150-5170.

Sandsmark, M., and Vennemo, H. (2007). A portfolio approach to climate investments: CAPM and endogenous risk. *Environmental and Resource Economics*, 37(4), 681-695.

Schlenker, W., and Roberts, D.L. (2009). Nonlinear temperature effects indicate severe damages to U.S. corn yields under climate change. *Proceedings of the National Academy of Sciences of the United States of America*, 106(37), 15594-15598.

Schlosser, C.A., Gao, X., Strzepek, K., Sokolov, A., Forest, C.E., Awadalla, S., and Farmer, W. (2012). Quantifying the likelihood of regional climate change: A hybridized approach. *Journal of Climate*, 26(10), 3394-3414.

Schlosser, A., Strzepek, K., Gao, X., Fant, C., Blanc, E., Paltsev, S., Jacoby, H., Reilly, J., and Gueneau, A. (2014). The future of global water stress: An integrated assessment. *Earth's Future*, 2(8), 341-361.

Schwinger, J., Tjiputra, J.F., Heinze, C., Bopp, L., Christian, J.R., Gehlen, M., Ilyina, T., Jones, C.D., Salas-Mélia, D., Segschneider, J., Séférian, R., and Totterdell, I. (2014). Nonlinearity of ocean carbon cycle feedbacks in CMIP5 earth system models. *Journal of Climate*, 27(11), 3869-3888.

Senior, C.A., and Mitchell, J.F. (2000). The time-dependence of climate sensitivity. *Geophysical Research Letters*, 27(17), 2685-2688.

Shindell, D.T. (2014). Inhomogeneous forcing and transient climate sensitivity. *Nature Climate Change*, 4, 274-277.

Slangen, A.B.A, Carson, M., Katsman, C.A., van de Wal, R.S.W., Köhl, A., Vermeersen, L.L.A., and Stammer, D. (2014). Projecting twenty-first century regional sea-level changes. *Climatic Change*, 124, 317-332.

Sohngen, B.L., Mendelsohn, R.O., and Sedjo, R.A. (2001). A global model of climate change impacts on timber markets. *Journal of Agricultural and Resource Economics*, 26(2), 326-343.

Solomon, S., Plattner, G., Knutti, R., and Friedlingstein, P., (2009). Irreversible climate change due to carbon dioxide emissions, *Proceedings of the National Academy of Sciences of the United States of America*, 106, 1704-1709.

Steinacher, M., Joos, F., Frolicher, T.L., Plattner, G.-K., and Doney, S.C. (2009). Imminent ocean acidification in the Arctic project with the NCAR global coupled carbon cycle-climate model. *Biogeosciences*, 6(4), 515-533.

Stern, N.H. (2007). *The Economics of Climate Change: The Stern Review*. Cambridge, U.K.: Cambridge University Press.

Sussman, F., Krishnan, N., Maher, K., Miller, R., Mack, C., Stewart, P., Shouse, K., and Perkins, W. (2014). Climate change adaptation cost in the U.S.: What do we know? *Climate Policy*, 14(2), 242-282.

Taheripour, F., Hertel, T.W., and Liu, J. (2013). The role of irrigation in determining the global land use impacts of biofuels. *Energy, Sustainability and Society*, 3(4), 1-18.

Tebaldi, C., and Arblaster, J.M. (2014). Pattern scaling: Its strengths and limitations, and an update on the latest model simulations. *Climatic Change*, 122(3), 459-471.

Tebaldi, C., and Knutti, R. (2007). The use of the multi-model ensemble in probabilistic climate projections. *Philosophical Transactions of the Royal Society A: Mathematical, Physical and Engineering Sciences*, 365(1857), 2053-2075.

Tokarska, K.B., Gillett, N.P., Weaver, A.J., Arora, V.K., and Eby, M. (2016). The climate response to five trillion tonnes of carbon. *Nature Climate Change*, 6, 851-855.

Tol, R.S.J. (2002a). Estimates of the damage costs of climate change—Part 1: Benchmark estimates. *Environmental and Resource Economics*, 21(1), 47-73.

Tol, R.S.J. (2002b). Estimates of the damage costs of climate change—Part II: Dynamic estimates. *Environmental and Resource Economics*, 21(2), 135-160.

Tol, R.S.J. (2009). The economic effects of climate change. *Journal of Economic Perspectives*, 23(2), 29-51.

Toya, H., and Skidmore, M. (2007). Economic development and the impact of natural disasters. *Economics Letters*, 94(1), 20-25.

Tsigas, M.E., Frisvold, G.B., and Kuhn, B. (1996). Global climate change in agriculture. In T.W. Hertel (Ed.), *Global Trade Analysis: Modelling and Applications* (pp. 280-304). New York: Cambridge University Press.

United Nations. (2004). *World Population to 2300*. New York: United Nations. Available: http://www.un.org/en/development/desa/population/publications/pdf/trends/WorldPop2300final.pdf [October 2016].

United Nations. (2015a). *World Population Prospects, The 2015 Revision: Key Conclusions and Advance Tables*. New York: United Nations. Available: http://esa.un.org/unpd/wpp/publications/files/key_Conclusions_wpp_2015.pdf [October 2016].

United Nations. (2015b). *World Population Prospects, The 2015 Revision: Methodology of the United Nations Population Estimates and Projections*. New York: United Nations.

U.S. Environmental Protection Agency. (2009). *Expert Elicitation Task Force White Paper. External Review Draft and Addendum: Selected Recent (2006-2008) Citations*. Washington, DC: U.S. Environmental Protection Agency.

U.S. Environmental Protection Agency. (2011). *The Benefits and Costs of the Clean Air Act from 1990 to 2020*. Washington, DC: U.S. Environmental Protection Agency.

U.S. Environmental Protection Agency. (2015). *Regulatory Impact Analysis for the Clean Power Plan Final Rule*. Available: https://www.epa.gov/sites/production/files/2015-08/documents/cpp-final-rule-ria.pdf [October 2016].

U.S. Government Accountability Office. (2014). *2014 Regulatory Impact Analysis: Development of Social Cost of Carbon Estimates*. Washington, DC: U.S. Government Accountability Office.

U.S. Office of Management and Budget. (1972). *Circular A-94: Discount Rates to Be Used in Evaluating Time-Distributed Costs and Benefits* (March 27). Washington, DC: Office of Management and Budget.

U.S. Office of Management and Budget. (2003). *Circular A-4: Regulatory Analysis* (September 17). Washington, DC: Office of Management and Budget.

van Vuuren, D.P., Edmonds, J., Kainuma, M., Riahi, K., Thomson, A., Hibbard, K., Hurtt, G.C., Kram, T., Krey, V., and Lamarque, J.-F. (2011). The representative concentration pathways: An overview. *Climatic Change, 109*, 1-4.

Vermeer, M., and Rahmstorf, S. (2009). Global sea level linked to global temperature. *Proceedings of the National Academy of Sciences of the United States of America, 106*(51), 21527-21532.

World Meteorological Organization. (2006). *Summary Statement on Tropical Cyclones and Climate Change*. Available: http://www.wmo.ch/pages/prog/arep/tmrp/documents/iwtc_summary.pdf [October 2016].

Waldhoff, S.T., Martinich, J., Sarofim, M., DeAngelo, B., McFarland, J., Jantarasami, L., Shouse, K., Crimmins, A., Ohrel, S., and Li, J. (2014). Overview of the special issue: A multi-model framework to achieve consistent evaluation of climate change impacts in the United States. *Climatic Change, 131*(1), 1-20.

Waldhoff, S.T., Martinich, J., Sarofim, M., DeAngelo, B., McFarland, J., Jantarasami, L., Shouse, K., Crimmins, A., Ohrel, S., and Li, J. (2015). Special issue on "A multi-model framework to achieve consistent evaluation of climate change impacts in the United States." *Climatic Change, 131*(1), 1-20.

Warren, R. (2011). The role of interactions in a world implementing adaptation and mitigation solutions to climate change. *Philosophical Transactions of the Royal Society of London A: Mathematical, Physical and Engineering Sciences, 369*(1934), 217-241.

Warren, R., Hope, C., Mastrandrea, M.D., Tol, R.S.J., Adger, N., and Lorenzoni, I. (2006). *Spotlighting Impacts in Integrated Assessment*. Tyndall Working Paper 91. Available: http://www.tyndall.ac.uk [October 2016].

Webster, M., Paltsev, S., Parsons, J., Reilly, J., and Jacoby, H. (2008). *Uncertainty in Greenhouse Gas Emissions and Costs of Atmospheric Stabilization*. Report No. 165. Cambridge, MA: Joint Program on the Science and Policy of Global Change. Available: http://globalchange.mit.edu/files/document/MITJPSPGC_Rpt165.pdf [October 2016].

Webster, M., Sokolof, A., Reilly, J., Forest, C., Paltsev, S., Schlosser, A., Wang, C., Kicklighter, D., Sarofim, S., Melillo, J., Prinn, R., and Jacoby, H. (2012). Analysis of climate policy targets under uncertainty. *Climatic Change, 112*(3), 569-583.

Weitzman, M.L. (2004). *Discounting a Distant Future Whose Technology Is Unknown.* Available: http://www.sv.uio.no/econ/english/research/news-and-events/events/guest-lectures-seminars/Thursday-seminar/2004/Thursday-spring04/weitzman-1.pdf [October 2016].

Weitzman, M.L. (2007). Subjective expectations and asset-return puzzles. *American Economic Review, 97*(4), 1102-1130.

Weitzman, M. (2011). Fat-tailed uncertainty in the economics of catastrophic climate change. *Review of Environmental and Economic Policy, 5*(2), 275-292.

Wilby, R.L., Dawson, C.W., and Barrow, E.M. (2002). SDSM: A decision support tool for the assessment of regional climate change impacts. *Environmental Modelling & Software 17*(2), 145-157.

Wise, M., Calvin, K., Thomson, A., Clarke, L., Bond-Lamberty, B., Sands, R., Smith, S.J., Janetos, A., and Edmonds, J. (2009). Implications of limiting CO_2 concentrations for land use and energy. *Science, 324*(5931), 1183-1186.

Yamamoto, A. Kawamiya, M., Ishida, A., Yamanaka, Y., and Watanabe, S. (2012). Impact of rapid sea-ice reduction in the Arctic Ocean on the rate of ocean acidification. *Biogeosciences, 9*(6), 2365-2375.

Yohe, G.W., and M.E. Schlesinger. (1998). Sea-level change: The expected economic cost of protection or abandonment in the United States. *Climatic Change, 38,* 337-342.

Yokohata, T., Emori, S., Nozawa, T., Ogura, T., Kawamiya, M., Tsushima, Y., Suzuki, T., Yukimoto, S., Abe-Ouchi, A., Hasumi, H., Sumi, A., and Kimoto, M. (2008). Comparison of equilibrium and transient responses to CO_2 increase in eight state-of-the-art climate models. *Tellus Series A Dynamic Meteorology and Oceanography, 60*(5), 946-961.

Yoshimori, M., Yokohata, T., and Abe-Ouchi, A. (2009). A comparison of climate feedback strength between CO_2 doubling and LGM experiments. *Journal of Climate, 22,* 3374-3395.

Zaveri, E., Grogan, D.S., Fisher-Vanden, K., Frolking, S., Lammers, R.B., Wrenn, D.H., Prusevich, A., and Nicholas, R.E. (2016). Invisible water, visible impact: Groundwater use and Indian agriculture under climate change. *Environmental Research Letters, 11*(8), 084005.

Zhou, Y., Eom, J., and Clarke, L. (2013). The effect of global climate change, population distribution, and climate mitigation on building energy use in the U.S. and China. *Climatic Change, 119*(3), 979-992.

Zickfeld, K., and Herrington, T. (2015). The time lag between a carbon dioxide emission and maximum warming increases with the size of the emission. *Environmental Research Letters, 10*(3), 031001.

Appendix A

Biographical Sketches of Committee Members and Staff

MAUREEN L. CROPPER (*Cochair*) is a distinguished university professor and chair of the Department of Economics at the University of Maryland. She is also a senior fellow at Resources for the Future and a research associate of the National Bureau of Economic Research. Previously, she was a lead economist at the World Bank. Her research has focused on valuing environmental amenities (especially environmental health effects), on the discounting of future health benefits, and on the tradeoffs implicit in environmental regulations. Her current research focuses on energy efficiency in India, on the impact of climate change on migration, and on the benefits of collective action in pandemic flu control. She has served as chair of the Economics Advisory Committee of the Science Advisory Board of the U.S. Environmental Protection Agency and as past president of the Association of Environmental and Resource Economists. She is a member of the National Academy of Sciences. She has a B.A. in economics from Bryn Mawr College and a Ph.D. in economics from Cornell University.

RICHARD G. NEWELL (*Cochair*) is the president and CEO of Resources for the Future. He is also an adjunct professor at Duke University, where he was previously the Gendell professor of energy and environmental economics and founding director of its Energy Initiative, and a research associate of the National Bureau of Economic Research. Previously, he held senior government positions, including as the administrator of the U.S. Energy Information Administration and as the senior economist for energy and environment on the President's Council of Economic Advis-

213

ers. He has published widely on the economics of markets and policies for energy and the environment, focusing on issues of global climate change, energy efficiency, and energy innovation. He is a member of the National Petroleum Council. He has an M.P.A. from the Woodrow Wilson School of Public and International Affairs at Princeton University and a Ph.D. from Harvard University.

MYLES ALLEN is professor of geosystem science in the Environmental Change Institute, the School of Geography and the Environment, and the Department of Physics, all at the University of Oxford. His research focuses on how human and natural influences on climate contribute to observed climate change and extreme weather events. He founded climateprediction.net and weatherathome.org experiments, using volunteer computing for weather and climate research. His recent work has dealt with quantifying the cumulative impact of carbon dioxide emissions on global temperatures and on the implications of reframing climate change as a carbon stock problem. He has served on several working groups on the physical science assessments of the Intergovernmental Panel on Climate Change and on the core writing team of the 2014 synthesis report. He is a recipient of the Appleton Medal from the Institute of Physics. He has a doctorate in physics from the University of Oxford.

MAXIMILIAN AUFFHAMMER is the George Pardee Jr. professor of international sustainable development and associate dean of social sciences at the University of California at Berkeley. His research focuses on environmental and resource economics, energy economics, and applied econometrics, and he has published widely on these topics. He is a research associate at the National Bureau of Economic Research in the Energy and Environmental Economics group, a Humboldt Fellow, and a lead author for the Intergovernmental Panel on Climate Change. He is a recipient of the Cozzarelli Prize awarded by the *Proceedings of the National Academies of Sciences* and of the Campus Distinguished Teaching Award and the Sarlo Distinguished Mentoring Award from the University of California, Berkeley. He has a B.S. in environmental science and an M.S. in environmental and resource economics from the University of Massachusetts at Amherst and a Ph.D. in economics from the University of California, San Diego.

CHRIS E. FOREST is associate professor of climate dynamics in the Department of Meteorology and Atmospheric Science and the Department of Geosciences, an associate in the Earth and Environmental Systems Institute, and associate director for the Network for Sustainable Climate Risk Management, all at Pennsylvania State University. He served as a lead author on the report of the Intergovernmental Panel on Climate Change for the chapter on the evaluation of climate models and on a report for the U.S. Climate Change Science Program examining the estimates of temperature trends in the atmospheric and surface climate data. He was elected to serve on the Electorate Nominating Committee for the Atmospheric and Hydrospheric Sciences Section of the American Association for the Advancement of Science. His research focuses on quantifying uncertainty in climate predictions and their implications for assessing climate risks. He has a B.S. in applied mathematics, engineering, and physics from the University of Wisconsin-Madison and a Ph.D. in meteorology from the Massachusetts Institute of Technology.

INEZ Y. FUNG is a professor of atmospheric science at the University of California, Berkeley. She is also a member of the science team for Orbiting Carbon Observatory of the U.S. National Aeronautics and Space Administration. She studies the interactions between climate change and biogeochemical cycles, particularly the processes that maintain and alter the composition of the atmosphere. Her research emphasis is on using atmospheric transport models and a coupled carbon-climate model to examine how CO_2 sources and sinks are changing. She is a recipient of the American Geophysical Union's Roger Revelle Medal. She is a member of the National Academy of Sciences, the American Academy of Arts and Sciences, and the American Philosophical Society. She is also a fellow of the American Meteorological Society and the American Geophysical Union. She has a B.S. in applied mathematics and a Ph.D. in meteorology from the Massachusetts Institute of Technology.

JAMES K. HAMMITT is professor of economics and decision sciences at the T.H. Chan School of Public Health and director of the Center for Risk Analysis, both at Harvard University, and an affiliate of the Toulouse School of Economics. His research concerns the development and application of quantitative methods—including benefit-cost, decision, and risk analysis—to health and environmental policy. Topics include management of long-term environmental issues with important scientific uncertainties, such as global climate change and stratospheric-ozone depletion, evaluation of ancillary benefits and countervailing risks associated with risk-control measures, and characterization of social preferences over health and environmental risks using revealed-preference, stated-preference, and health-utility methods. He has a Ph.D. in public policy from Harvard University.

HENRY D. JACOBY is the William F. Pounds professor of management (emeritus) in the Sloan School of Management and a founding co-director of the Joint Program on the Science and Policy of Global Change, both at the Massachusetts Institute of Technology (MIT). Previously, he served on the faculties of the Department of Economics and the Kennedy School of Government at Harvard University. His work has been devoted to integration of the natural and social sciences and policy analysis in application to issues of energy and environment, with a recent focus on the threat of global climate change. He has been director of the Harvard Environmental Systems Program, director of the MIT Center for Energy and Environmental Policy Research, associate director of the MIT Energy Laboratory, and chair of the MIT faculty. Other activities have included service on the National Petroleum Council, the Nuclear Fuels Working Group of the Atlantic Council and the Scientific Committee of the International Geosphere-Biosphere Program, and he was a convening lead author of the 2014 National Climate Assessment. He has an undergraduate degree in mechanical engineering from the University of Texas at Austin and a Ph.D. in economics from Harvard University.

JENNIFER A. HEIMBERG (*Study Director*) is a senior program officer in the Division of Earth and Life Studies (DELS) and the Division of Behavioral and Social Sciences and Education. In her work for the Nuclear and Radiation Studies Board in DELS, she has focused on nuclear security, nuclear detection capabilities, and environmental management issues, and she has directed studies and workshops related to nuclear proliferation, nuclear terrorism, and the management of nuclear wastes. Previously, she worked as a program manager at the Johns Hopkins University Applied Physics Laboratory, where she established its nuclear security program with the Department of Homeland Security's Domestic Nuclear Detection Office. She has a B.S. in physics from Georgetown University, a B.S.E.E. from Catholic University, and a Ph.D. in physics from Northwestern University.

ROBERT E. KOPP is associate director of the Energy Institute and an associate professor in the Department of Earth & Planetary Sciences at Rutgers University. His research focuses on understanding uncertainty in past and future climate change, with major emphases on sea level change and on the interactions between physical climate change and the economy. He was a contributing author to the working groups on physical science and on impacts, adaptation, and vulnerability of the fifth assessment report of the Intergovernmental Panel on Climate Change. Previously, he served at the U.S. Department of Energy as a science and technology policy fellow of the American Association for the Advancement of science and as a postdoctoral fellow in geosciences and public policy at Princeton University. He is a past Leopold leadership fellow and a recipient of the

Sir Nicholas Shackleton Medal of the International Union for Quaternary Research and the William Gilbert Medal of the American Geophysical Union. He has an undergraduate degree in geophysical sciences from the University of Chicago and a Ph.D. in geobiology from the California Institute of Technology.

WILLIAM PIZER is a professor in the Sanford School of Public Policy at Duke University. His current research examines how public policies to promote clean energy can effectively leverage private-sector investments, how environmental regulation and climate policy can affect production costs and competitiveness, and how the design of market-based environmental policies can be improved. He has published widely on these topics. Previously, he was Deputy Assistant Secretary for Environment and Energy at the U.S. Department of the Treasury, overseeing the department's role in the domestic and international environment and energy agenda of the United States. He also previously was a researcher at Resources for the Future in Washington, D.C. He has a bachelor's degree in physics from the University of North Caroling at Chapel Hill and a master's degree and a Ph.D. in economics from Harvard University.

STEVEN K. ROSE is a senior research economist in the Energy and Environmental Research Group at the Electric Power Research Institute. His research focuses on long-term modeling of energy systems and climate change drivers, mitigation, and potential risks, as well as the economics of land use and bioenergy as they relate to climate change and energy policy. His work also considers climate change risks and responses, trade-offs between mitigation and temperature, the marginal costs of climate change, energy-water-land linkages, the role of bioenergy in long-term climate management, the economics of REDD+ and agricultural productivity, and mitigation institutions, investment risks and incentives. He was a lead author for the Fifth and Fourth Assessment Reports of the Intergovernmental Panel on Climate Change and of the U.S. National Climate Assessment. He also serves on the federal government's U.S. Carbon Cycle Science Program Carbon Cycle Scientific Steering Group and a panel of Science Advisory Board of the U.S. Environmental Protection Agency. He has a B.A. in economics from the University of Wisconsin-Madison and a Ph.D. in economics from Cornell University.

RICHARD SCHMALENSEE is the Howard W. Johnson professor of management (emeritus) and professor of economics (emeritus) at the Massachusetts Institute of Technology (MIT). Previously at MIT, he was the John C. Head III dean of the Sloan School of Management, director of the Center for Energy and Environmental Policy Research, and a member of the Energy Council. He also previously served as a member of the President's Council of Economic Advisers. His research and teaching have focused on industrial organization economics and its applications to business decision making and public policy. He is a fellow of the Econometric Society and of the American Academy of Arts and Sciences. He has served on the executive committee of the American Economic Association and as a director of several corporations, and he is currently chair of the board of Resources for the Future. He is a distinguished fellow of the Industrial Organization Society. He has an S.B. and a Ph.D. in economics from the Massachusetts Institute of Technology.

JOHN P. WEYANT is professor of management science and engineering, director of the Energy Modeling Forum, and deputy director of the Precourt Institute for Energy Efficiency, all at Stanford University. His current research focuses on analysis of global climate change policy options, energy efficiency analysis, energy technology assessment, and models for strategic planning. He has been a convening lead author or lead author for the several chapters of the Intergovernmental Panel on Climate Change (IPCC) reports, and, most recently, as a review editor for the climate change mitigation working group of the IPCC's fourth and fifth assessment reports. He was also a founder and serves as chair of the Scientific Steering Committee of the Integrated Assessment Modeling Consortium, an collaboration of 53 member institutions from around the world. He is a recipient of the Adelmann-Frankel award from the U.S. Association for Energy Economics for unique and innovative contributions to the field of energy economics. He has a B.S./M.S. in aeronautical engineering and astronautics and M.S. degrees in engineering management and in operations research and statistics from Rensselaer Polytechnic Institute, and a Ph.D. in management science from University of California, Berkeley.

CASEY J. WICHMAN (*Technical Consultant*) is a fellow at Resource for the Future in Washington, D.C. His research is concentrated at the intersection of environmental and public economics, with an emphasis on examining the ways in which individuals make decisions in response to environmental policies using quasi-experimental techniques. In particular, his work analyzes the effectiveness of price and non-price interventions for water conservation, the role of information in the design of environmental policy, and the effect of water scarcity in the energy sector. He has a B.A. in economics from Ithaca College, an M.S. in economics from North Carolina State University, and an M.S. and a Ph.D. in agricultural and resource economics from the University of Maryland, College Park.

Appendix B

Presentations to the Committee

Washington, DC, September 2, 2015

- *Sponsors' Interests and Goals for the Study*: Kenneth Gillingham, Council of Economic Advisers
- *Methodology for the Social Cost of Carbon Estimates*: Elizabeth Kopits, U.S. Environmental Protection Agency

Washington, DC, November 13, 2015

- *Damage Models for Existing Integrated Assessment Models (IAMs) and Areas of Research*
  - Improving the damage portion of IAMs DICE model: Kenneth Gillingham and William Nordhaus, Yale University
  - FUND model: David Anthoff, University of California, Berkeley
  - PAGE model: Chris Hope, University of Cambridge
- *Current State of Evidence and Approaches, Options for Integration into IAMs*: Leon Clarke, Pacific Northwest National Laboratory, and Sol Hsiang, University of California, Berkeley
- *Current State of Evidence and Approaches, Future Research Needs*: John Reilly, Massachusetts Institute of Technology, and Wolfram Schlenker, Columbia University

Washington, DC, May 5, 2016

- *Inter-Sectoral Impact Model Intercomparison Project (ISI-MIP) Overview*: Katja Frieler, lead of the ISI-MIP project, Potsdam Institute of Climate Impact Research, Germany
- *Costs of Perturbations and Feedbacks in the CO_2 and Methane Cycles*: David Archer, University of Chicago
- Market and Nonmarket Damages Panel
  - *A New Empirical Approach to Global Damage Function Estimation*: Michael Greenstone, University of Chicago
  - *Nonmarket Damages from Climate Change*: Michael Hanemann, Arizona State University

Washington, DC, August 23, 2016

- Remarks and Q&A with several members of the Interagency Working Group on the Social Cost of Greenhouse Gases: Elizabeth Kopits, U.S. Environmental Protection Agency; Alex Marten, U.S. Environmental Protection Agency; Elke Hodson, U.S. Department of Energy; Sheila Olmstead, Council of Economic Advisers

Appendix C

Elicitation of Expert Opinion

Expert judgment about the value of a model parameter or other quantity can be obtained using many methods (Morgan and Henrion, 1990; O'Hagan et al., 2006; U.S. Environmental Protection Agency, 2009). The terms "expert elicitation" and "structured expert elicitation" (or "structured expert judgment") are typically used to describe a formal process in which multiple experts report their individual subjective probability distributions for the quantity. This usage is distinct from less formal methods in which someone provides a best guess or other estimate of the quantity. In practice, experts may provide a full probability distribution or a few fractiles of a distribution (often, three or five).

Expert elicitation can be distinguished from other formal methods of collecting experts' judgments, such as group processes. These processes include expert committees (like those of the National Academies of Sciences, Engineering, and Medicine or the Intergovernmental Panel on Climate Change), in which experts reach a consensus through loosely structured or unstructured interaction, and more structured group processes such as the Delphi method (Dalkey, 1970; Linstone and Turoff, 1975), in which experts provide a probability distribution or other response, receive information on other experts' responses (without associating individuals with responses), provide a revised response, and iterate until the process converges.

An important concern with expert elicitation is that subject-matter experts (like most other people) have little experience or skill in reporting their beliefs in the form of a subjective probability distribution. Their judgments about probabilities and other quantities are often consistent with the hypothesis that they are influenced by cognitive heuristics that lead to systematic biases (as elucidated by Tversky and Kahneman, 1974). Also, to be most useful, subjective probability distributions must be well calibrated. Unfortunately, many experts (and others) prove to be over-confident in that they provide probability distributions that are too narrow, that is, the true or realized values are too frequently in the tails of their estimated distributions.

The hypothesis that an expert is well calibrated can be tested if the expert provides distributions for multiple quantities for which the values can be known, so one can determine the fractile of the corresponding subjective distribution at which each true or realized value falls. Calibration means that the realizations are consistent with the hypothesis that they are random draws from the experts' corresponding distributions. For example, the expected fraction of realizations that fall outside the ranges defined by the expert's 10th and 90th percentiles for the corresponding quantities is 20 percent; the probability that the actual number of realizations outside these intervals could have arisen by chance if the expert were well calibrated can be calculated using conventional statistical methods.

Expect elicitation can be conducted using more or less elaborate methods. Many practitioners use an elaborate approach (Morgan and Henrion, 1990; Evans et al., 1994; Budnitz et al., 1997; O'Hagan et al., 2006; Goossens et al., 2008; U.S. Environmental Protection Agency, 2009; Morgan, 2014). Knol et al. (2010) outline a seven step procedure for organizing an expert elicitation, illustrated in Figure C-1.

Expert elicitation is often conducted by a study team consisting of an interviewer and a subject-matter expert (who may be assisted by other personnel). The study team meets with and interviews each expert individually. The interviewer is familiar with issues of subjective probability and appropriate methods for eliciting judgments but may have little familiarity with the subject about which judgments are elicited, while the subject-matter expert is familiar with the topics about which judgments are elicited and the available theory, measurements, and other evidence from which the experts may form their judgments.

An expert-elicitation study often includes three distinct steps. First, the study team collects a set of relevant background studies that are provided to each expert, to make sure each is familiar with and can easily consult the relevant literature.

Second, there is an in-person meeting of the study team and all the selected experts, at which the experts discuss and if necessary clarify the definition of the quantities about which their judgments will be elicited; discuss the strengths and weaknesses of available empirical studies and

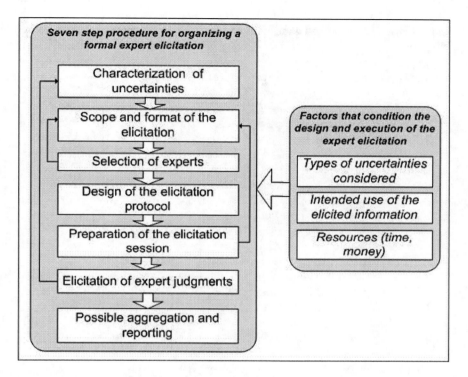

FIGURE C-1 Steps for a formal expert elicitation.
SOURCE: Knol et al. (2010, Figure 1).

other evidence; are familiarized with the elicitation procedures; and are introduced to concepts of subjective probability and common cognitive heuristics and biases (such as overconfidence). Such preparation may also include practice judgments about quantities whose values are subsequently revealed.

The third step is in-person interviews with each expert, during which the expert provides subjective probability distributions for the relevant quantities. These interviews often take several hours. The interviewer is primarily concerned with the procedure, framing questions to elicit the expert's best judgment, minimizing effects of cognitive heuristics, question order, and other factors. The subject-matter expert is more concerned with the expert's responses and rationales and can query the expert about the basis for the stated distributions, pursue the lines of evidence she or he finds more and less persuasive, and explore the extent to which the expert has incorporated evidence that seems to conflict with the stated distribution. In some cases, either the expert or the study team may draft an explanation of the expert's rationale, which the expert is expected to endorse (after revision, as appropriate). The expert may be invited to revise her or his responses after the interview, if desired.

During the elicitation process, it is common to help the expert address the quantity from multiple perspectives, to help in reporting her or his best judgment. For example, the same concept could be framed alternatively as a growth rate or a growth factor (or a future level conditional on a specified current level). If the expert provides a distribution about the growth rate, the study team could convert this to a distribution for the growth factor and allow the expert to contemplate whether this distribution is compatible with her or his beliefs or if the distribution for the growth rate needs to be adjusted. (Alternatively, the distribution for the growth factor could be elicited and the implied distribution of the growth rate derived from it.)

Similarly, the study team can help the expert view the estimated distribution from multiple perspectives, allowing adjustment until comfortable with the result. For example, the procedure could first ask for the median, described as the value such that the true quantity is equally likely to be larger or smaller. Then the expert could be asked, "If you learned the true value was larger than the median, what value do you judge it equally likely that the true value is above or below (i.e., what is the upper quartile of the distribution)?" The lower quartile can be elicited by an analogous question, after which the team can ask whether the expert believes it equally likely the true value is inside or outside the interquartile range.

An alternative to this series of questions is to begin by asking the expert for some extreme fractiles (e.g., the 10th and 90th percentiles) and then the more central fractiles such as the median. An advantage of eliciting fractiles in the tails (before the center) is to help protect against the problem of overconfidence that can arise from beginning with a central value and then adjusting away from it, but not sufficiently far, consistent with the "anchoring and adjustment" heuristic of Tversky and Kahneman (1974). To help the expert report extreme fractiles, the team might ask the expert to describe conditions under which the quantity would be larger than the expert's largest (or smaller than the expert's smallest) fractile; thinking about these conditions may induce the expert to revise these fractiles.

As an alternative to this elaborate, in-person interview process, elicitations can be conducted through telephone, email, or survey methods. A disadvantage of these less intensive methods is that the study team has less ability to help the expert think carefully about multiple lines of evidence and to view the issue from multiple perspectives. There seems to be little evidence from which one can judge the effects of alternative elicitation procedures on the quality of the resulting distributions, because investigators have not often elicited distributions for the same or similar quantities using alternative procedures.

As alluded to in the above example of growth rate or growth factor, it is not always clear what quantity is best to elicit. When several quantities are logically related, it is probably best to encourage the expert to think carefully about all of them: at a minimum, it may be helpful for the expert to evaluate the implications of the offered distribution for one quantity on the implied distribution for the others. A related question is the degree of disaggregation: the process could elicit an aggregate quantity, the components from which it can be calculated, or both. The best approach may be the one with which each expert is most comfortable. If distributions for components are elicited, constructing the implied distribution for the aggregate requires information on the conditionality between the components, that is, the distribution for one component conditional on (in principle) all possible realizations of the other components. If the expert provides a distribution for only the aggregate, in principle she or he must take account of this conditionality implicitly.

One criterion for choosing the quantity to be elicited is that it ought to be a quantity for which the true value can be, at least in principle, probably by some form of measurement. This criterion implies the quantity is sufficiently well defined to remove any ambiguity about what would be measured. It is sometimes described as a clairvoyance test, meaning that a clairvoyant would be able to report the true quantity (without requiring any clarification). In contrast, an abstract model parameter may not be suitable if the true value of that parameter depends on the assumption that the model is accurate, particularly if the expert rejects that assumption.

Another important question is how to select experts. Given the burden of elicitation, it may be too difficult to recruit a large number. Many studies use between 5 and 15; there is some evidence of sharply diminishing returns beyond about 10 (Hora, 2004). Typically, the experts who are sought span the range of defensible perspectives about a quantity, but it is not necessary or appropriate to have the distribution of experts match the population frequency of alternative views (within the expert community). Commonly advocated methods of expert selection include inviting people whose work is most often cited or asking such people whom they would nominate as well qualified. In general, the set of experts who provide judgments is made public, but the matching between individual experts and distributions is concealed. The rationale for this approach is that it allows experts freedom to provide their best judgments without concern for representing the position of an employer or other party.

After subjective probability distributions are elicited from multiple experts, the question remains how to use them (which is related to the question of which experts to select). At a minimum, it seems useful to report the distribution provided by each expert, so readers have some appreciation for the degree of homo- or heterogeneity among the responses. It may be useful to understand the reasons for large differences among experts' distributions: for example, experts may differ significantly in their interpretation of the credibility or relevance of particular data or theories. Beyond this reporting, it seems useful to combine the distributions using either an algorithmic approach or (possibly) a social or judgmental approach. Some elicitation experts (Keith, 1996; Morgan, 2014) have argued not to combine the distributions of multiple experts, but rather that the overall analysis ought to be replicated using each expert's judgments individually as input to the evaluation. To the extent the overall conclusion is insensitive to which expert's distributions are used, this approach may be adequate; if the conclusions depend on the expert, one can either report the multiple conclusions that result from using each expert's judgments individually or find some way to combine them. If the evaluation yields a probability distribution of some output (e.g., the SC-CO$_2$), one could combine the multiple output distributions that result from using each expert's distributions for the inputs using an algorithmic or other approach. It would be interesting to compare the properties of combining the experts' distributions to use as input or conducting the analysis using each expert's distributions alone then combining the output distributions.

A number of algorithmic methods for combining experts' distributions have been studied. In principle, a Bayesian approach in which the experts' distributions are interpreted as data and used to update some prior distribution seems logical, but it is problematic. Such an approach requires a joint likelihood function, that is, a joint conditional distribution that describes the probability that each expert will provide each possible subjective distribution, conditional on the true value of the quantity. This distribution encapsulates information about the relative quality of the experts and about their dependence, which may be difficult to obtain and to evaluate.

The most common approaches to combining experts' distributions are a simple or weighted average. The simple average is often used because it seems fair and avoids treating experts differently. The notion of eliciting weights (from the experts about themselves or about the other experts) has been considered.

Cooke (1991) has developed a performance-weighted average, which has been applied in many contexts (Goossens et al., 2008). The weights depend on experts' performance on "seed" quantities, which are quantities whose value becomes known after the experts' provide distributions for them. Performance is defined as a combination of calibration and informativeness, where informativeness is a measure of how concentrated (narrow) the distribution is. Clearly the judgments of a well-calibrated expert who provides narrow distributions are more valuable than the judgments of an expert who provides poorly calibrated or uninformative distributions. A key question is whether one can identify seed quantities that have the property that one would put more weight on the judgment of an expert for the quantity of interest when that expert provides better calibrated and more informative distributions for the seed variables. Cooke and Goossens (2008) have shown that the performance-weighted average of distributions usually outperforms the simple average, where performance is again measured again by calibration and informativeness (and is often evaluated on seed variables not used to define the weights, because the value of the quantity of interest in many expert elicitation studies remains unknown). Some authors remain skeptical, however (e.g., Morgan, 2014). The simple average distribution may be reasonably well calibrated, but it tends to be much less informative than the performance-weighted combination.

Note that when taking a linear combination of experts' judgments, such as a simple or weighted average, it is desirable to average the probabilities, not the fractiles. Averaging the fractiles is equivalent to taking the harmonic mean of the probabilities, and hence it tends to yield very low probabilities on values for which any expert provided a small probability and to concentrate the distribution on values to which all experts assign relatively high probability (Bamber et al., 2016). Using the harmonic-mean probability is likely to accentuate the problem of overconfidence (distributions that are too narrow).

Expert elicitation is a method for characterizing what is known about a quantity; it does not add new information as an experiment or measurement would. Ideally, it captures the best judgments of the people who have the most information and deepest understanding of the quantity of interest. For some quantities, there may be so little understanding of the factors that affect its magnitude that informed judgment is impossible or can produce only uselessly wide bounds. For these quantities, neither expert elicitation nor any alternative can overcome the limits of current knowledge. Only additional research can push back those limits.

REFERENCES

Bamber, J.L., Aspinall, W.P., and Cooke, R.M. (2016). A commentary on "How to interpret expert judgment assessments of twenty-first century sea-level rise" by Hylke de Vries and Roderik SW van der Wal. *Climatic Change, 137*, 321.

Budnitz, R.J., Apostolakis, G., Boore, D.M., Cluff, L.S., Coppersmith, K.J., Cornell, C.A., and Morris, R.A. (1997). *Recommendations for Probabilistic Seismic Hazard Analysis: Guidance on Uncertainty and Use of Experts.* NUREG/CR-6372, Vol. 2. Washington, DC: U.S. Nuclear Regulatory Commission.

Cooke, R.M. (1991). *Experts in Uncertainty: Opinion and Subjective Probability in Science.* New York: Oxford University Press.

Cooke, R.M., and Goossens, L.H.J. (2008). TU Delft expert judgment data base. *Reliability Engineering and System Safety, 93,* 657-674.

Dalkey, N.C. (1970). *The Delphi Method: An Experimental Study of Group Opinion.* Technical report RM-5888-PR. Santa Monica, CA: RAND Corporation.

Evans, J.S., Gray, G.M., Sielken, R.L., A.E. Smith, A.E., Valdez-Flores, C., and Graham, J.D. (1994). Use of probabilistic expert judgment in uncertainty analysis of carcinogenic potency. *Regulatory Toxicology and Pharmacology, 20,* 15-36.

Goossens, L.H.J., Cooke, R.M., Hale, A.R., Rodic'-Wiersma, Lj. (2008). Fifteen years of expert judgement at TUDelft. *Safety Science, 46,* 234-244.

Hora, S.C. (2004). Probability judgments for continuous quantities: Linear combinations and calibration. *Management Science, 50,* 567-604.

Keith, D.W. (1996). When is it appropriate to combine expert judgments? *Climatic Change, 33,* 139-143.

Knol, A.B., Slottje, P., van der Sluijs, J.P., and Lebret, E. (2010). The use of expert elicitation in environmental health impact assessment: A seven step procedure. *Environmental Health, 9,* 19.

Linstone, H., and Turoff, M. (1975). *The Delphi Method: Techniques and Applications,* Reading, MA: Addison-Wesley.

Morgan, M.G. (2014). Use (and abuse) of expert elicitation in support of decision making for public policy. *Proceedings of the National Academy of Sciences of the United States of America, 111,* 7176-7184.

Morgan, M.G., and Henrion, M. (1990). *Uncertainty: A Guide to Dealing with Uncertainty in Quantitative Risk and Policy Analysis.* Cambridge, U.K.: Cambridge University Press.

O'Hagan, A., Buck, C.E., Daneshkhah, A., Eiser, J.R., Garthwaite, P.H., Jenkinson, D.J., Oakley, J.E., and Rakow, R. (2006). *Uncertain Judgements: Eliciting Experts' Probabilities.* Hoboken, NJ: Wiley.

Tversky, A., and Kahneman, D. (1974). Judgment under uncertainty: Heuristics and biases, *Science, 185,* 1124-1131.

U.S. Environmental Protection Agency. (2009). *Expert Elicitation Task Force White Paper: External Review Draft and Addendum: Selected Recent (2006-2008) Citations.* Science Policy Council. Washington, DC: U.S. Environmental Protection Agency.

Appendix D

Global Growth Data and Projections

In support of the committee's recommendation for estimating a probability density of average annual growth rates of global per capita gross domestic product (GDP) using global data (see Chapter 3, Recommendation 3-2), this appendix describes the Mueller-Watson (2016) approach (hereafter, MW) as a demonstration of how the recommendations could be followed. It details the data source and implementation of the MW approach, along with the results.

DATA

As described below one could construct two time series for economic growth based on the Maddison Project Database.[1] The Maddison Project provides lengthy time series of per capita income for virtually all countries. Its starting point was the seminal work of Summers and Heston (1984), updated in the Penn World Tables, on real GDP in purchasing power parity terms (via the Geary–Khamis method) for all countries since 1950; Maddison obtained corresponding population data from the United Nations. Additional countries and years were obtained through review and compilation of individual country estimates from a wide range of

---

[1]The database is based on the work of the economic historian Angus Maddison and is freely available from the Maddison Project (http://www.ggdc.net/maddison/maddison-project/home.htm [November 2016]). The Conference Board is currently responsible for maintaining the data (currently referred to as the Total Economy Database) and has updated the series since 2010.

economic historians; they were initially published in book form (some released through the Organization for Economic Cooperation and Development [OECD]). Since 2010, a small group of scholars have collaborated to carry on this work.

Despite the unprecedented coverage and availability of the Maddison data, the length of available data varies by country and is missing for some years. Consistent coverage begins later for less developed countries or developed countries whose economies were adversely affected by World War II. Hence, there is a tradeoff between coverage and the length of the series.

When forecasting global growth over a time horizon of several centuries, the optimal tradeoff between coverage and timespan is not obvious. Without arguing in favor of any particular sample, two are considered in this example. Focusing on the post-1950 time period, all countries are available to estimate average annual growth rate for the world for 60 years. This forms the basis of the first time series. The basis of the second time series is a panel of 25 countries—which accounted for as much as 63 percent of global GDP in 1950 but as little as 46 percent of global GDP in 2009—that are available from 1870, thus providing data for 140 years. Those 25 countries are Australia, Austria, Belgium, Brazil, Canada, Chile, Denmark, Finland, France, Greece, Germany, Italy, Japan, Netherlands, New Zealand, Norway, Portugal, Spain, Sri Lanka, Sweden, Switzerland, United Kingdom, United States, Uruguay, and Venezuela.

The selection of 1870 as the starting year seemed to be the best compromise between breadth and depth. Prior to 1870, annual data are not available for 12 of the 25 countries: Austria, Brazil, Canada, Chile, Greece, Japan, New Zealand, Portugal, Spain, Sri Lanka, Uruguay, and Venezuela. Shortening the time series by 50 years would add only 7 more countries: Argentina (starting in 1875), India (starting in 1884), Mexico (starting in 1900), Ecuador (starting in 1900), Ireland (starting in 1921), Turkey (starting in 1923), and South Africa (starting in 1924). Because these 25 countries tend to be more developed, they have a slower average growth rate (1.73%) than the average growth for all countries in the world (2.19%) for the 1950-2010 period.

To apply the MW approach to the Maddison data, one must construct a univariate series for global growth rates. World GDP per capita is already aggregated and provided directly by the Maddison Project for 1950-2010. To construct the second series, the population tables provided by in the original Maddison data (through 2009) can be used to convert GDP per capita to GDP, which can be aggregated and then divided by the aggregate population of the 25 countries in this example. The growth rate is constructed by taking the first difference of the logs of aggregate GDP per capita. The resulting growth rates are shown in Table D-1 and in Figure D-1, which also displays the results of filtering out the short-run variation in the raw data. Figure D-1 also shows the long-run variation of the growth rates (i.e., with frequency less than $q\pi/T$). The estimated predictive density is shown in Figure D-2. Summary statistics for this distribution are given in Table D-2.

TABLE D-1 Growth Rates of Aggregate GDP per Capita, in percent

Year	25 Countries 1870-2010	Entire World 1950-2010
1871	1.58	NA
1872	2.93	NA
1873	1.04	NA
1874	2.40	NA
1875	1.94	NA
1876	−2.04	NA
1877	1.10	NA
1878	0.80	NA
1879	0.64	NA
1880	4.66	NA
1881	1.98	NA
1882	2.34	NA
1883	1.27	NA
1884	0.11	NA
1885	−0.24	NA
1886	1.04	NA
1887	2.34	NA
1888	0.50	NA
1889	2.36	NA
1890	0.97	NA
1891	0.42	NA
1892	2.48	NA
1893	−1.48	NA
1894	1.09	NA
1895	3.55	NA
1896	0.06	NA
1897	2.31	NA
1898	3.13	NA
1899	3.23	NA
1900	0.80	NA
1901	2.15	NA
1902	0.07	NA
1903	1.81	NA
1904	−0.20	NA
1905	2.50	NA
1906	5.13	NA
1907	1.45	NA
1908	−3.95	NA
1909	4.13	NA
1910	0.20	NA
1911	2.77	NA
1912	2.72	NA

continued

Year	25 Countries 1870-2010	Entire World 1950-2010
1913	1.87	NA
1914	−7.78	NA
1915	0.69	NA
1916	6.61	NA
1917	−2.84	NA
1918	0.49	NA
1919	−1.13	NA
1920	0.71	NA
1921	−1.17	NA
1922	5.78	NA
1923	3.55	NA
1924	4.18	NA
1925	3.00	NA
1926	2.10	NA
1927	2.23	NA
1928	2.34	NA
1929	3.13	NA
1930	−6.07	NA
1931	−7.12	NA
1932	−6.81	NA
1933	1.24	NA
1934	4.33	NA
1935	3.88	NA
1936	6.04	NA
1937	3.98	NA
1938	−0.14	NA
1939	5.89	NA
1940	1.65	NA
1941	6.61	NA
1942	6.52	NA
1943	7.83	NA
1944	1.70	NA
1945	−10.27	NA
1946	−13.33	NA
1947	1.05	NA
1948	3.93	NA
1949	2.61	NA
1950	5.67	NA
1951	5.35	4.08
1952	2.72	2.65
1953	3.51	3.05
1954	1.27	1.42
1955	5.30	4.32
1956	2.26	2.70
1957	2.24	1.69
1958	0.10	1.12
1959	4.60	2.62
1960	3.70	3.65
1961	2.89	2.05
1962	4.09	2.84

TABLE D-1 Continued

Year	25 Countries 1870-2010	Entire World 1950-2010
1963	3.22	2.14
1964	4.83	5.02
1965	3.88	3.16
1966	4.27	3.30
1967	2.63	1.65
1968	4.56	3.28
1969	4.23	3.35
1970	2.67	3.07
1971	2.43	1.91
1972	4.09	2.69
1973	4.97	4.52
1974	0.17	0.39
1975	−0.59	−0.26
1976	3.85	3.06
1977	2.70	2.24
1978	3.28	2.60
1979	2.95	1.74
1980	0.47	0.25
1981	0.44	0.26
1982	−1.01	−0.50
1983	1.52	0.86
1984	3.81	2.79
1985	2.79	1.70
1986	2.45	1.77
1987	2.52	2.04
1988	3.29	2.49
1989	2.52	1.51
1990	0.73	0.32
1991	0.22	−0.16
1992	0.98	0.42
1993	0.49	0.74
1994	2.32	1.97
1995	1.84	2.58
1996	1.87	1.88
1997	2.70	2.51
1998	1.88	0.52
1999	2.27	2.30
2000	2.93	3.47
2001	0.62	1.70
2002	0.62	2.29
2003	1.09	3.47
2004	2.40	3.85
2005	1.78	3.18
2006	2.06	3.85
2007	1.87	3.08
2008	−0.48	1.62
2009	−4.31	−1.96
2010	NA	4.39

NOTES: NA, not available. See text for explanation of the calculation.

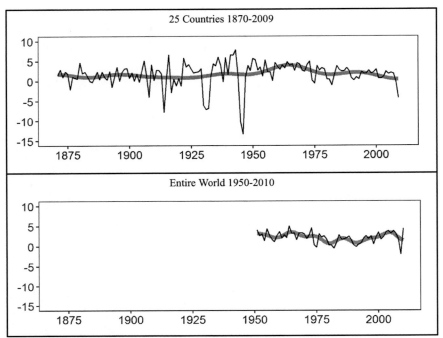

FIGURE D-1 Observed growth rates and their long-run variation.
NOTE: See text for discussion.

IMPLEMENTATION

The MATLAB code for implementing the MW approach is freely available from Mark Watson's website.[2] Only a subset of the code is required to generate the results presented below: lr_main_annual.m, figure_1_2.m, Sigma_Compute.m, den_invariate.m, psi_compute.m, t_mixture.m, and lr_pred_set.m. It is possible to replicate our estimates using the nine-step procedure detailed in the rest of this section.[3]

Step One

Alter the directory paths and file names in the code to point to the data.

---

[2]See https://www.princeton.edu/~mwatson/wp.html [November 2016].
[3]Note: these calculations were initially implemented in R but replicated in MATLAB using the MW files.

Step Two

Generate the $q = 12$ cosine transformations (the MW recommendation) and project the growth time series x of length T onto the space spanned by a constant (which picks up the unconditional mean of the series) and a set of q regressors that are cosine transformations of the data to isolate the low frequency variation in the data, denoted hereafter as the $q \times 1$ vector X. Plot as depicted in Figure D-1 (above) (see the figure_1_2.m script).

Although a value of $q = T$ would preserve all of the information in the original time series, MW recommend trimming it to $q = 12$. Truncating the set at $q < T$ does involve some loss of information and thus some loss of econometric efficiency; a larger q would decrease the uncertainty in the predictions of growth rates. However, a larger q weakens the approximations utilized by this approach: the distribution of the transformed data would be further from the limiting normality and the shape of spectrum could exhibit greater deviations from the approximate shape near a frequency of 0 (the latter of which is not mitigated by a larger sample size T). According to the numerical calculations of MW, a value of $q = 12$ tends to optimize the tradeoff between efficiency and robustness.

Step Three

Change the forecasting horizon(s) to the desired number of years (e.g., 90 = 2100-2010 or 290 = 2300-2010 in this application). Note that the available data for this example ends in 2009 or 2010, depending on the dataset, which is the year that the forecast begins.

Step Four

Specify the prior for the order of integration of the time series data generating process, denoted as d, on the near-0 spectrum by setting $b = c = 0$ in equation (20) of MW, which is the simpler prior that it discusses.

Step Five

Compute the $q + 1$ dimensional covariance matrix Σ for each d (using the scripts Sigma_Compute.m and associated subroutines):

$$\begin{bmatrix} \Sigma_{XX}(d) & \Sigma_{XY}(d) \\ \Sigma_{YX}(d) & \Sigma_{YY}(d) \end{bmatrix}.$$

The particular value of d (along with $q = 12$ and the forecast horizon h/T) is a critical input into the computation of each Σ term, making each Σ term a complicated function of d as detailed in MW, Appendix A-4.

TABLE D-2 Summary Statistics of Uncertainty Distributions of Average Annual Growth Rates

Horizon	25 Countries 1870-2009		World 1950-2010	
	2010-2100 (90 Years)	2010-2300 (290 Years)	2010-2100 (90 Years)	2010-2300 (290 Years)
Mean	1.37	1.44	2.14	2.18
Std. Deviation	1.02	1.34	1.03	1.40
1st Percentile	−1.85	−3.08	−1.17	−2.42
5th Percentile	−0.44	−0.80	0.56	0.29
10th Percentile	0.15	0.06	1.13	1.07
25th Percentile	0.90	0.99	1.75	1.78
33rd Percentile	1.13	1.23	1.91	1.95
50th Percentile	1.49	1.58	·2.18	2.20
66th Percentile	1.78	1.85	2.42	2.45
75th Percentile	1.96	2.04	2.60	2.64
90th Percentile	2.42	2.63	3.13	3.31
95th Percentile	2.78	3.20	3.59	3.99
99th Percentile	3.71	4.89	4.95	6.26

The unobserved random variable Y_T is the average growth rate from time $T + 1$ to time $T + h$, relative to the observed average growth rate from $t = 1$ to T:

$$Y_T = \bar{x}_{T+1:T+h} - \bar{x}_{1:T}.$$

Conditional on d, the following statistic is distributed as a Student's t with $q = 12$ degrees of freedom (see MW, equation (8)):

$$\frac{\left[Y_T / \sqrt{X'X}\right] - \Sigma_{YX}\Sigma_{XX}^{-1}\left[X / \sqrt{X'X}\right]}{\sqrt{\Sigma_{YY} - \Sigma_{YX}\Sigma_{XX}^{-1}\Sigma_{XY}} \times \sqrt{\left[X / \sqrt{X'X}\right]'\Sigma_{XX}^{-1}\left[X / \sqrt{X'X}\right]/q}} \Bigg| d \sim Student's\ t_{q=12},$$

where the explicit dependence of each Σ term on d has been suppressed for the sake of notational brevity, mimicking MW. Note that $\Sigma_{YX}\Sigma_{XX}^{-1}X$ is the mean predicted value of Y_T for each value of d, as implied by the symmetry of the Student's t-distribution.

Step Six

Compute the likelihood and posterior for d along a grid of n values, assuming a U[-0.4,1.0] prior following MW. One can then compute a predictive density for Y_T over a grid of values, averaging the conditional density based on the above Student's t using the posterior for d (see the lr_main_annual.m script).

Step Seven

Plot predictive distribution as Figure D-2 (see the figure_1_2.m script).

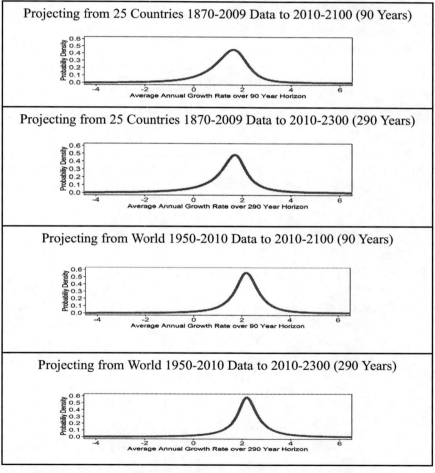

FIGURE D-2 Predictive distribution for average annual growth rates.
NOTE: See text for discussion.

Step Eight

Compute the summary statistics of the predictive distribution, shown in Table D-2, above. To further simplify notation, let $m(d)$ and $s(d)$ be such that:

$$m(d) = \Sigma_{YX}(d)\Sigma_{XX}^{-1}(d)X$$
$$s(d) = \sqrt{\Sigma_{YY}(d) - \Sigma_{YX}(d)\Sigma_{XX}^{-1}(d)\Sigma_{XY}(d)} \times \sqrt{X'\Sigma_{XX}^{-1}(d)X/q}.$$

The mean growth rate can be computed by weighting the conditional means $m(d)$ by the posterior for d, then adding to $\bar{x}_{1:T}$. To compute the percentiles, first substitute these variables into the above distributional result, making it clear that the distribution of Y_T given d can be written in terms of a Student's t:

$$\frac{Y_T - m(d)}{s(d)} \Big| d \sim Student's\ t_{q=12}.$$

Note a $\sqrt{X'X}$ has been cancelled in both the numerator and the denominator of the earlier expression. The unconditional cumulative distribution function of Y_T, that is, not conditional on d, is then given by the finite weighted sum of Student's t-distributions:

$$G(Y_T) = \sum_{i=1}^{n} Pr(d = d_i)F\left(\frac{Y_T - m(d)}{s(d)}\right),$$

where F is the cumulative distribution function (CDF) of the Student's t with $q = 12$ degrees of freedom.

The percentiles appearing in Table D-2 (above) of $G(Y_T)$ can then be computed directly from the replication code associated with MW, using the t_mixture.m script. This script numerically inverts $G(Y_T)$.

Step Nine

As discussed in Chapter 3, one could approximate the distribution of Y_T using three equally weighted values based on the tercile means. To compute the means of the terciles of average growth rate from time $T + 1$ to time $T + h$, shown in Table D-3, one would have to go beyond the MW analysis. Each tercile is defined as a range: from the 0th percentile (negative infinity) to the 33rd percentile is the lower tercile, from the 33rd percentile to 66th percentile is the middle tercile, and from the 66th percentile

TABLE D-3 Mean Growth Rate Conditional on Tercile of Uncertainty Distribution

	25 Countries 1870-2009		World 1950-2010	
Horizon	2010-2100 (90 Years)	2010-2300 (290 Years)	2010-2100 (90 Years)	2010-2300 (290 Years)
Bottom Tercile	0.31	0.18	1.19	1.04
Middle Tercile	1.47	1.56	2.17	2.20
Top Tercile	2.33	2.57	3.07	3.31

to the 100th percentile (positive infinity) is the top tercile. The expectation of the average growth rate from time $T + 1$ to $T + h$, conditional on the average growth rate falling in range $R = (\alpha, \beta)$, is

$$E\left[\bar{x}_{T+1:T+h} \middle| \bar{x}_{T+1:T+h} \in R\right] = \bar{x}_{1:T} + E\left[Y_T \middle| \alpha < (Y_T + \bar{x}_{1:T}) < \beta\right].$$

Substituting in the definition for the conditional expectation of Y_T:

$$E\left[\bar{x}_{T+1:T+h} \middle| \bar{x}_{T+1:T+h} \in R\right] = \bar{x}_{1:T} + \int_{\alpha-\bar{x}_{1:T}}^{\beta-\bar{x}_{1:T}} \frac{Y_T g(Y_T)}{G(\beta-\bar{x}_{1:T}) - G(\alpha-\bar{x}_{1:T})} dY_T.$$

where $g()$ is the pdf corresponding to $G()$. By virtue of α and β being the range of a tercile, the difference in the denominator will be equal to $1/3$. Substituting in the full expression for the mixture of densities for $g()$:

$$E\left[\bar{x}_{T+1:T+h} \middle| \bar{x}_{T+1:T+h} \in R\right] = \bar{x}_{1:T} + \sum_{i=1}^{n} \frac{\Pr(d=d_i)}{3} \times \left[\int_{\alpha-\bar{x}_{1:T}}^{\beta-\bar{x}_{1:T}} \frac{Y_T}{s(d_i)} f\left(\frac{Y_T - m(d_i)}{s(d_i)}\right) dY_T\right],$$

where $f()$ is the probability density of a Student's t with $q = 12$ degrees of freedom.

Each term inside of the summation operator can then be scaled by the probability of being within the given tercile, conditional on d,

$$F\left(\frac{\beta-\bar{x}_{1:T} - m(d_i)}{s(d_i)}\right) - F\left(\frac{\alpha-\bar{x}_{1:T} - m(d_i)}{s(d_i)}\right).$$

This is not necessarily $1/3$: for some values of d, this will be more and for others, less. This allows us to include the corresponding reciprocal of this factor inside the integral:

$$E\left[\bar{x}_{T+1:T+h}\middle|\bar{x}_{T+1:T+h} \in R\right] = \bar{x}_{1:T} +$$

$$\sum_{i=1}^{n} \Pr(d = d_i) \frac{F\left(\dfrac{\beta - \bar{x}_{1:T} - m(d_i)}{s(d_i)}\right) - F\left(\dfrac{\alpha - \bar{x}_{1:T} - m(d_i)}{s(d_i)}\right)}{3}$$

$$\times \frac{\displaystyle\int_{\alpha - \bar{x}_{1:T}}^{\beta - \bar{x}_{1:T}} \frac{Y_T}{s(d_i)} f\left(\dfrac{Y_T - m(d_i)}{s(d_i)}\right) dY_T}{F\left(\dfrac{\beta - \bar{x}_{1:T} - m(d_i)}{s(d_i)}\right) - F\left(\dfrac{\alpha - \bar{x}_{1:T} - m(d_i)}{s(d_i)}\right)},$$

where, again, F is the CDF for a Student's t with $q = 12$ degrees of freedom. With a change of variables $Z_T = (Y_T - m)/s$, the last term becomes:

$$m(d_i) + s(d_i) \left[\frac{\displaystyle\int_{\frac{\alpha - \bar{x}_{1:T} - m(d_i)}{s(d_i)}}^{\frac{\beta - \bar{x}_{1:T} - m(d_i)}{s(d_i)}} Z_T f(Z_T) dZ_T}{F\left(\dfrac{\beta - \bar{x}_{1:T} - m(d_i)}{s(d_i)}\right) - F\left(\dfrac{\alpha - \bar{x}_{1:T} - m(d_i)}{s(d_i)}\right)} \right].$$

Notice that the bracketed term is just the expectation of a random variable Z_T distributed as a standard Student's t with $q = 12$ degrees of freedom, falling in the range

$$\left(\frac{\alpha - \bar{x}_{1:T} - m(d_i)}{s(d_i)}, \frac{\beta - \bar{x}_{1:T} - m(d_i)}{s(d_i)}\right).$$

This term can be computed in closed form in terms of gamma functions and the CDF of the standard Student's t-distribution with $q = 12$ degrees of freedom, which one can obtain from the work on truncated t-distributions by Kim (2008, p. 84):

$$E\big(Z\big|Z\in(a,b)\big)=\frac{\Gamma\left(\dfrac{q-1}{2}\right)q^{\frac{q}{2}}}{2\big[F(b)-F(a)\big]\Gamma\left(\dfrac{q}{2}\right)\Gamma\left(\dfrac{1}{2}\right)}\left(\big(q+a^{2}\big)^{\frac{-q-1}{2}}-\big(q+b^{2}\big)^{\frac{-q-1}{2}}\right).$$

REFERENCES

Kim, H.-J. (2008). Moments of truncated Student-t distribution. *Journal of the Korean Statistical Society, 37*, 81–87.

Mueller, U.K., and Watson, M.W. (2016). Measuring uncertainty about long-run predictions. *Review of Economic Studies, 83*(4), 1711-1740.

Summers, R., and Heston, A. (1984). Improved international comparisons of real product and its composition: 1950–1980. *Review of Income and Wealth, 30*(2), 207-219.

Appendix E

Comparison of a Simple Earth System Model to Existing SC-IAMs

This appendix compares the climate components in the existing integrated assessment models used to produce estimates of the SC-CO_2 (SC-IAMs) to those of the Finite Amplitude Impulse Response (FAIR) model, the illustrative simple Earth system model described in Chapter 4: see Tables E-1 and E-2. In Table E-1, the shaded rows indicate the top-level description of the component of the simple Earth system model. The clear rows are descriptions of important "response characteristics" of the components (i.e., timescales and feedbacks).

The three SC-IAMs differ substantively from FAIR, as well as from each other, in such characteristics as the structure and response timescales of the global climate and carbon cycle and the modeling of the concentrations and forcing imparted by non-CO_2 greenhouse gases and aerosols. As shown in Table E-2, the SC-IAM climate component modeling differs in other characteristics as well: regional climate, the way in which the CO_2 pulse is implemented, treatment of parametric uncertainty, and time steps. As discussed in Chapter 4, IWG updates would need to consider how each of the characteristics in Table E-2 will be handled in future modeling.

As shown by Rose et al. (2014), these differences across SC-IAMs affect the reference climate projections and CO_2 pulse responses by producing significant differences in future concentrations and global average warming by 2100 for the same emissions inputs, as well as differences in the timing, magnitude, and shape of incremental temperature responses to a CO_2 pulse: see Figure E-1. Future research ought to consider a similar

243

TABLE E-1 Global Climate and Carbon Cycles of SC-IAMs and FAIR

Element	DICE 2010	FUND 3.8	PAGE09	FAIR
Climate				
Structure	Two-box model (surface land/ocean and deep ocean)	One-timescale impulse response function	One-timescale impulse response function	Two-timescale impulse response function
Global Mean Temperature Adjustment Timescales	Function of equilibrium climate sensitivity (ECS): impulse response equivalent for ECS = 3 has slow response of > 200 years, fast response of ~30 years.	Function of ECS: for ECS of 3 °C and default parameters, *e*-folding time* 44 years; increases as quadratic function of ECS	Constant: Modal half-life of 35 years [*e*-folding time of 50 years]	For 3 °C ECS, median slow response timescale of 249 years, fast response of 4.1 years (Geoffroy et al., 2013). Response coefficients adjusted to prescribed values or distributions of ECS and TCR (Millar et al., 2015)
Non-CO_2 Forcings	Exogenous forcing must be prescribed.	CH_4, N_2O, SF_6, and SO_2 modeled with single timescales; no other non-CO_2 forcers	Exogenous well-mixed greenhouse gas forcing only	Exogenous forcing must be prescribed, or adjustment timescales and radiative efficacies can be adjusted to represent non-CO_2 greenhouse gases (Myhre et al., 2013).

TABLE E-1 Continued

Element	DICE 2010	FUND 3.8	PAGE09	FAIR
Carbon Cycle				
Structure	Three-box model (atmosphere, surface ocean, deep ocean)	Five-timescale impulse response function	Three-timescale impulse response function	Four-timescale impulse response function
Timescales	Fraction 1: > 50 years Fraction 2: > 1,000 years Fraction 3: infinite lifetime [inferred from inspection]	Fraction 1: 10%, 2 years Fraction 2: 25%, 17 years Fraction 3: 32%, 74 years Fraction 4: 20%, 363 years Fraction 5: 13%, infinite lifetime	For modal values, Fraction 1: 40%, zero lifetime Fraction 2: 25%, 123 years Fraction 3: 35%, infinite lifetime	For modal values, Fraction 1: 27%, 4 years Fraction 2: 28%, 35 years Fraction 3: 23%, 381 years Fraction 4: 22%, infinite lifetime (Myhre et al., 2013)
Carbon Cycle Feedback	None	Terrestrial carbon stock loss with warming (with central parameter values: ~0.14% of terrestrial carbon stock in a given period released per degree of warming relative to 2010)	Atmospheric CO_2 increase with warming (with central parameter values: 10% CO_2 concentration gain per period per °C, with maximum of 50%)	Airborne fraction increases as a linear function of warming and cumulative land and ocean carbon uptake (Millar et al., 2016)

* "*e*-folding time" is the time-scale for exponential decay to an equilibrium state.

TABLE E-2 Additional Characteristics of the Climate Components
of the SC-IAMs

Element	DICE 2010	FUND 3.8	PAGE09
Climate			
Regional Temperatures	None	Pattern scaling for 14 regions based on 14 general circulation models (Gates et al., 1996, as cited in Mendelsohn et al., 2000)	Parameterized downscaling based on Intergovernmental Panel on Climate Change (2007) that scales an assumed temperature difference between equator and pole by regional latitude and average land/ocean warming ratio
Ocean			
Global Mean Sea Level Rise (GMSL)	Equilibrium for components (thermal expansion, glacier melt, GIS^a mass loss, WAIS^b mass loss) computed as function of temperature; adjustment time exogenous for thermal expansion and glaciers, and function of temperature for GIS and WAIS	Equilibrium GMSL rise computed as a function of temperature; exogenous adjustment time	Equilibrium GMSL rise computed as a function of temperature; exogenous adjustment time
Regional Sea Level Rise	Identical to global mean	Identical to global mean	Identical to global mean
Ocean pH	Not calculated	Not calculated	Not calculated

TABLE E-2 Continued

Element	DICE 2010	FUND 3.8	PAGE09
Numerical Implementation			
Model Time Step	10 years	1 year	Variable: 10-year 2000-2060, 20-year 2060-2100, and 100-year 2100-2300
Implementation of CO_2 Pulse in Year t	Pulse spread equally over the decade straddling year t	Pulse spread equally over the decade from year t forward	Pulse distributed evenly over the two decades preceding and subsequent to year t
Parametric Uncertainty Included (Other Than ECS)	No	Yes	Yes

[a]Greenland ice sheet.
[b]West Antarctic ice sheet.

comparison to FAIR, with comparisons of deterministic and probabilistic behavior. (See below for discussion of SC-IAM parametric uncertainty.)

The SC-IAM models also vary notably in their sensitivity to alternative assumptions explored in the current IWG approach, such as emissions and equilibrium climate sensitivity (ECS), with the Framework for Uncertainty, Negotiation, and Distribution (FUND) model being less responsive than the Dynamic Integrated Climate-Economy (DICE) and Policy Analysis of the Greenhouse Effect (PAGE) models to different emissions scenarios and ECS values (Rose et al., 2014). This reduced insensitivity to varying ECS in FUND arises because, by construction, the response timescales are adjusted automatically as ECS varies to account for the inverse correlation between the rate of temperature response and ECS. DICE also modifies the temperature adjustment timescale with ECS, while PAGE makes no countervailing adjustment and, therefore, is more responsive to ECS. Marten (2013) characterizes FUND's temperature adjustment timescale response as somewhat ad hoc.

Many of the differences between the SC-IAM climate models do not represent structural uncertainty, that is, different representations of the underlying system dynamics, which is the primary motivation for using multiple models. The climate component of all three SC-IAMs can be interpreted as special cases of FAIR, with the differences between them resulting from parameter choices or the setting of certain parameters to zero. These choices generate different response behaviors, the significance

(a) CO_2 concentrations

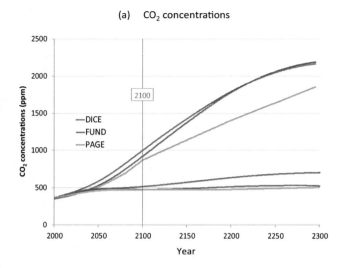

(b) Global mean temperatures above preindustrial levels

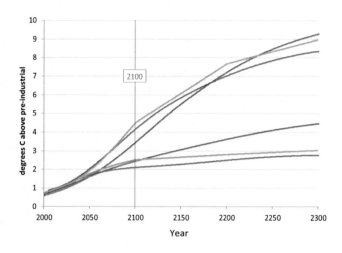

FIGURE E-1 Plots of CO_2 concentrations and global mean temperatures generated from diagnostic tests of the SC-IAM climate models with high (solid) and low (dashed) emissions scenarios (a) and (b): CO_2 concentrations and global mean temperatures above preindustrial levels to the year 2300. (c) and (d): incremental increases in CO_2 concentration and global mean temperature to the year 2300 from a CO_2 emissions pulse in 2020.

NOTES: The diagnostics were run with the IWG high and low greenhouse gas emissions scenarios (Interagency Working Group, 2010, 2013a, 2013b, 2015, 2016),

(c) Incremental CO$_2$ concentration increase

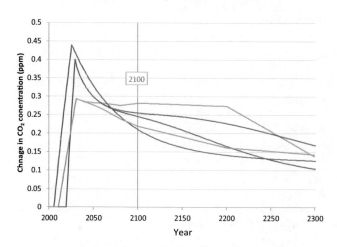

(d) Incremental global mean temperature increase

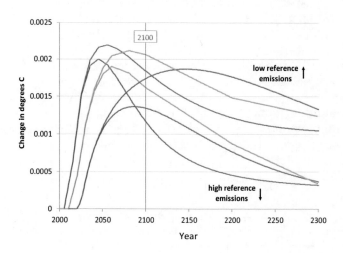

and the incremental responses are from a 1 billion metric ton carbon (3.67 billion metric ton CO$_2$) emissions pulse experiments applied to the respective high and low reference emissions (pulse released in year 2020 only). Fossil fuel and industrial CO$_2$ emissions in the high IWG scenario are 67, 118, 144, and 102 Gt CO$_2$/year in 2050, 2100, 2200, and 2300, respectively. In the IWG low emissions scenario, they are 22, 23, 14, and 7 Gt CO$_2$/year, respectively. For comparison, RCP 8.5 fossil fuel and industrial CO$_2$ emissions are 74, 105, 56, and 7 Gt CO$_2$/year in 2050, 2100, 2200, and 2300, respectively; and, in RCP 2.6 are 12, –3, –3, and –3 Gt CO$_2$/yr.
SOURCE: Rose et al. (2014).

of which could be quantified by future analyses. DICE omits all feedbacks between climate and the carbon cycle (equivalent to $r_T = 0$ and $r_C = 0$ in FAIR). PAGE and FUND both omit the second timescale in the climate response ($d_2 = 0$), and they only explicitly include the impact of warming on airborne fraction ($r_C = 0$). All the models exclude some categories of non-CO_2 forcings. These differences produce variations in projected climate variables between the SC-IAMs that should not be interpreted as representing scientific uncertainty.

The SC-IAM components are also implemented differently in the IWG modeling in terms of their treatment of uncertainty and CO_2 pulse implementation. On uncertainty, FUND and PAGE include parametric uncertainty in certain model-specific variables; DICE does not. Also, the parametric uncertainties in FUND and PAGE are specified very differently, such that PAGE generates a significantly larger uncertainty range and produces higher average warming than FUND (Rose et al., 2014). For pulse implementation, there are differences in the timing of the incremental CO_2 concentration and temperature responses due to how a CO_2 pulse is introduced into and propagates through each model. Like the parameter choices discussed above, the exclusion of parametric uncertainty from DICE and the differences in pulse implementation contribute to variations in results across models that artificially represent actual scientific uncertainty.

In summary, the climate models incorporated in DICE, FUND, and PAGE are structurally equivalent to special cases of FAIR: although all omit at least one key element, they could be modified to be equivalent to FAIR and thus to satisfy the criteria outlined in Recommendation 4-1 and the requirements in Conclusion 4-1, in Chapter 4. The chapter also covers how the implementation differences discussed above could be addressed.

REFERENCES

Gates, W.L., Henderson-Sellers, A., Boer, C., Folland, B. Kitoh, A., McAvaney, F., Semazzi, N., Smith, A., Weaver, Q., and Zeng, C. (1996). Climate models—evaluation. In J.T. Houghton, L.G. Meiro Filho, B.A. Callander, N. Harris, A. Kattenburg, K. Maskell (Eds.), *Climate Change 1995: The Science of Climate Change* (pp. 229-284). Cambridge, U.K.: Cambridge University Press.

Geoffroy, O., Saint-Martin, D., Bellon, G., Voldoire, A., Olivié, D.J.L., and Tytéca, S. (2013). Transient climate response in a two-layer energy-balance model. Part II: Representation of the efficacy of deep-ocean heat uptake and validation for CMIP5 AOGCMs. *Journal of Climate, 26*(6), 1859-1876.

Intergovernmental Panel on Climate Change. (2007). *Climate Change 2007: Working Group I: The Physical Science Basis. Contribution to the Fourth Assessment Report of the Intergovernmental Panel on Climate Change.* S. Solomon, D. Qin, M. Manning, Z. Chen, M. Marquis, K.B. Averyt, M. Tignor, and H.L. Miller (Eds). Cambridge, U.K. and New York: Cambridge University Press.

Interagency Working Group on the Social Cost of Carbon. (2010). *Technical Support Document: Social Cost of Carbon for Regulatory Impact Analysis Under Executive Order 12866* (February 2010). Available: https://obamawhitehouse.archives.gov/sites/default/files/omb/inforeg/for-agencies/Social-Cost-of-Carbon-for-RIA.pdf [January 2017].

Interagency Working Group on the Social Cost of Carbon. (2013a). *Technical Support Document: Technical Update of the Social Cost of Carbon for Regulatory Impact Analysis Under Executive Order 12866* (May 2013). Available: https://obamawhitehouse.archives.gov/sites/default/files/omb/inforeg/social_cost_of_carbon_for_ria_2013_update.pdf [January 2017].

Interagency Working Group on the Social Cost of Carbon. (2013b). *Technical Support Document: Technical Update of the Social Cost of Carbon for Regulatory Impact Analysis Under Executive Order 12866* (November 2013 Revision). Available: https://obamawhitehouse.archives.gov/sites/default/files/omb/assets/inforeg/technical-update-social-cost-of-carbon-for-regulator-impact-analysis.pdf [January 2017].

Interagency Working Group on the Social Cost of Carbon. (2015). *Technical Support Document: Technical Update of the Social Cost of Carbon for Regulatory Impact Analysis Under Executive Order 12866* (July 2015 Revision). Available: https://obamawhitehouse.archives.gov/sites/default/files/omb/inforeg/scc-tsd-final-july-2015.pdf [January 2017].

Interagency Working Group on the Social Cost of Greenhouse Gases. (2016). *Technical Support Document: Technical Update of the Social Cost of Carbon for Regulatory Impact Analysis Under Executive Order 12866* (September 2016 Revision). Washington, DC: Interagency Working Group on the Social Cost of Carbon. Available: https://obamawhitehouse.archives.gov/sites/default/files/omb/inforeg/august_2016_sc_ch4_sc_n2o_addendum_final_8_26_16.pdf [January 2017].

Marten, A.L., Kopp, R.E., Shouse, K.C., Griffiths, C.W., Hodson, E.L., Kopits, E., Mignone, B.K., Moore, C., Newbold, S.C., Waldhoff, S., and Wolverton, A. (2013). Improving the assessment and valuation of climate change impacts for policy and regulatory analysis. *Climatic Change, 117*(3), 433-438.

Mendelsohn, R., Schlesinger, M., and Williams, L. (2000). Comparing climate impacts across models. *Integrated Assessment, 1*(1), 37.

Millar, R.J., Otto, A., Forster, P.M., Lowe, J.A., Ingram, W.J., and Allen, M.R. (2015). Model structure in observational constraints on transient climate response. *Climatic Change, 131*(2), 199-211.

Millar, R.J., Nicholls, Z.R., Friedlingstein, P., and Allen, M.R. (2016). A modified impulse-response representation of the global response to carbon dioxide emissions. *Atmospheric Chemistry and Physics*, 1-20.

Myhre, G., Shindell, D., Bréon, F.M., Collins, W., Fuglestvedt, J., Huang, J., Koch, D., Lamarque, J.F., Lee, D., Mendoza, B., Nakajima, T., Robock, A., Stephens, G., Takemura, T., and Zhang, H. (2013). Anthropogenic and natural radiative forcing. In T.F. Stocker, D. Qin, G.-K. Plattner, M. Tignor, S.K. Allen, J. Boschung, A. Nauels, Y. Xia, V. Bex, and P.M. Midgley (Eds.), *Climate Change 2013: The Physical Science Basis. Working Group I Contribution to the Fifth Assessment Report of the Intergovernmental Panel on Climate Change* (pp. 659-740). Cambridge, U.K. and New York: Cambridge University Press.

Rose, S., Turner, D., Blanford, G., Bistline, J., de la Chesnaye, F., and Wilson, T. (2014). *Understanding the Social Cost of Carbon: A Technical Assessment*. Report 3002004657. Palo Alto, CA: Electric Power Research Institute.

Appendix F

Empirical Equation for Estimating Ocean Acidification

This appendix provides the supporting material for estimating the pH of seawater that may be required for the climate change damage estimates discussed in Chapter 5. It covers two approaches consistent with the simple Earth system model detailed in Chapter 4. The first approach estimates globally averaged pH directly from globally averaged atmospheric CO_2. The second approach estimates pH from surface temperature and the ocean carbon concentrations and may be useful for estimating regional changes in pH or global changes in pH in SC-CO_2 models with interactive ocean carbon modules. In both approaches, regression relationships are derived from outputs from a full ocean carbonate chemistry code run for typical ranges of ocean seawater temperatures, dissolved inorganic carbon concentrations, and chemical compositions. This allows applying the standard set of equations to both global and regional estimates of pH.

OVERVIEW

Carbonate chemistry in the ocean comprises mainly two reversible reactions:

$$CO_2 + H_2O \leftrightarrow H^+ + HCO_3^-$$

(1)

$$H^+ + CO_3^{2-} \leftrightarrow HCO_3^-.$$

(2)

The reactions are governed by the 1st and 2nd dissociation constants of carbonic acid, both of which vary with temperature and salinity:

$$K_1 = \frac{[H^+][HCO_3^-]}{[CO_2]} \qquad (3)$$

$$K_2 = \frac{[H^+][CO_3^{2-}]}{[HCO_3^-]}. \qquad (4)$$

The reversible reactions (1) and (2) must conserve mass of dissolved inorganic carbon (DIC) and charge. Charge balance is represented by total alkalinity (A_T), which is the number of moles of hydrogen ion equivalent to the excess of proton acceptors over proton donors in 1 kg of sample (Dickson, 1981):

$$DIC = [CO_2] + [HCO_3^-] + [CO_3^-] \qquad (5)$$

$$A_T = [HCO_3^-] + 2[CO_3^{2-}] + [B(OH)_4^-] + [OH^-] + [HPO_4^{2-}] + 2[PO_4^{3-}] + \ldots (6)$$

For given values of DIC and A_T, equations (3), (4), (5), and (6) are four equations in four unknowns and can be solved iteratively.

To derive the empirical relationships describing the dependence of pH on temperature and carbon concentration, the publicly available carbonate chemistry code CO2SYS.m is used (van Heuven et al., 2011). Temperature-dependent solubility is from Weiss (1974); equilibrium constants K_1 and K_2 are from Luecker et al. (2000); and those for the species in A_T (boric acid, hydrogen fluoride, phosphoric acid, and silicic acid) are from Dickson et al. (2007). The globally averaged A_T for the upper 100m is 2311 microeq/kg seawater, based on the GLODAP2 gridded data (Key et al., 2004). The concentration of borates varies linearly with salinity, and it is 415.7 micromol/kg for the Luecker et al. (2000) equilibrium constants. Other standard ocean values are S = 35 psu, and the concentrations of silicate and phosphate are 50 and 2 micromol/kg, respectively. The code CO2SYS.m was run for concentrations of DIC ranging from 1800 to 2300 micromol/kg seawater, and for surface temperatures ranging from 0 to 50°C. Outputs pH and the partial pressure of CO_2 in surface water (pCO$_2$) are used for regression analyses below.

APPROACH 1: GLOBALLY AVERAGED PH

The concentration of hydrogen ion is directly related to the concentration of CO_2 solution (see Equation [1], above), and a simple relationship between pH and the partial pressure of pCO$_2$ can be derived:

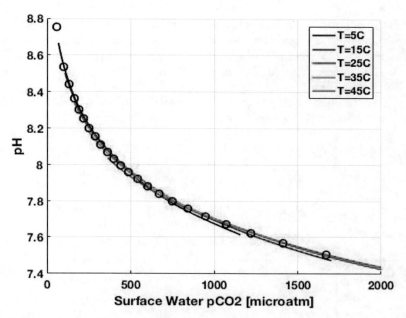

FIGURE F-1 Variation of pH with the partial pressure of CO_2 in surface waters as calculated with CO2SYS.m for different temperatures (solid lines) and with Equation (7) (circles).

$$pH = -0.3671 \cdot \log_e(pCO_2) + 10.2328, \tag{7}$$

where $pH = -\log_{10}[H^+]$ is defined on the "total" hydrogen ion scale (Dickson et al., 2007) and pCO_2 is in micro-atmospheres.[1] Figure F-1 shows that pH estimated by Equation (7) closely matches that calculated by CO2SYS.m for various temperatures. Temperature dependence has been subsumed into the determination of pCO_2.

Globally averaged pCO_2 of the surface ocean can be estimated from globally averaged CO_2 concentration in the atmosphere with approximately 1 year lag.

---

[1]An atmospheric CO_2 concentration of 400 ppm (10^{-6} mol CO_2 per mol air) is equivalent to an atmospheric CO_2 partial pressure of 400 microatm.

APPROACH 2: REGIONAL PH

Temperature, DIC, and hence pCO_2 and pH of the surface ocean vary from place to place and from season to season. Results from CO2SYS.m show the variations of pH as a function of DIC and temperature: see Figure F-2. As can be seen, pH decreases with increasing dissolved inorganic carbon and with increases temperature.

The committee derived empirical fits to the results shown in Figure F-2. DIC is in micromol/kg seawater, and T is temperature in Celsius.

$$pH = p1(T)*DIC*DIC + p2(T)*DIC + p3(T) \qquad (8)$$

with

$$p1(T) = q1(1)*T*T + q1(2)*T + q1(3)$$
$$p2(T) = q2(1)*T*T + q2(2)*T + q2(3)$$
$$p3(T) = q3(1)*T*T + q3(2)*T + q3(3).$$

	i = 1	i = 2	i = 3
q1(i)	1.32165e-10	1.52051e-08	−2.37923e-06
q2(i)	−4.82195e-07	-5.89841e-05	7.69483e-03
q3(i)	4.59338e-04	4.05966e-02	2.58590e+00

The pH values calculated using Equation (7) are shown as circles in Figure F-1.

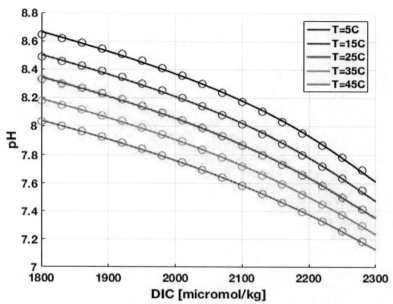

FIGURE F-2 Variation of pH with DIC and temperature, as calculated with carbon chemistry code CO2SYS.m (solid) and with Equation (8) (circles).

ESTIMATION OF DIC INCREASE

These calculations assume[2] the preindustrial near-surface ocean has $DIC_0 = 2005$ micromol/kg and $T_0 = 15°C$. For an increase of X Pg CO_2 in the upper ocean box of volume V (m³), the change in DIC can be estimated from:

$$DIC = DIC_0 + \Delta DIC \tag{9}$$

$$\Delta DIC = X \cdot \lambda \text{ where } \lambda = \frac{MW}{\rho \cdot V}. \tag{10}$$

In Equation (10), $MW = 10^{21}/44$ micromol/PgCO₂ is the molecular weight of CO_2 and $\rho = 1024$ kg/m³ is the density of seawater. For a 100 m deep global ocean box, $\lambda = 0.634$. For alternative upper-ocean volumes, λ will require a recalibration to be consistent with the transient climate response and equilibrium climate sensitivity estimates that implicitly include a heat capacity estimate as given in FAIR or the SC-IAMs.

---

[2]See http://www.whoi.edu/OCB-OA/page.do?pid=112136 [November 2016].

REFERENCES

Dickson, A.G. (1981). An exact definition of total alkalinity and a procedure for the estimation of alkalinity and total inorganic carbon from titration data. *Deep Sea Research, 28A,* 609-623.

Dickson, A.G., Sabine, C.L. and Christian, J.R. (Eds.). (2007). *Guide to Best Practices for Ocean Co_2 Measurements.* PICES Special Publication 3. Sidney, British Columbia: North Pacific Marine Science Organization.

Key, R.M., Kozyr, A., Sabine, C.L., Lee, K, Wanninkhof, R., Bullister, J., Feely, R.A., Millero, F. Mordy, C., and Peng. T.-H. (2004). A global ocean carbon climatology: Results from GLODAP. *Global Biogeochemical Cycles, 18,* GB4031. Available: http://cdiac.ornl.gov/ftp/oceans/GLODAP_Gridded_Data [November 2016].

Lueker, T.J., Dickson, A.G., and Keeling, C.D. 2000. Ocean pCO_2 calculated from dissolved inorganic carbon, alkalinity, and equations for K1 and K2: Validation based on laboratory measurements of CO_2 in gas and seawater at equilibrium. *Marine Chemistry, 70,* 105-119.

Van Heuven, S., Pierrot, D., Rae, J.W.B., Lewis, E., and Wallace, D.W.R. (2011). *MATLAB Program Developed for CO_2 System Calculations.* ORNL/CDIAC-105b. Oak Ridge, TN: U.S. Department of Energy.

Weiss, R.F. (1974). Carbon dioxide in water and seawater: The solubility of a non-ideal gas. *Marine Chemistry, 2,* 203-215.

Appendix G

Damages Model-Specific Improvement Opportunities

In this appendix, the committee suggests model-specific improvements that could be undertaken if the IWG chooses to continue to use all or a subset of the current SC-IAM damage formulations. Based on our review and understanding of the current SC-IAM damage formulations, opportunities for updating each SC-IAM to satisfy the criteria in Recommendation 2-2 (in Chapter 2) have been identified. The committee's goal is to highlight opportunities for the IWG to consider as alternatives in its decision process for implementing a near-term update. The committee views the existing models as providing material that is readily available, pieces of which can be used and updated and combined with other pieces, to create an improved damages module in the near term.

DICE

If in the near term the IWG decides to continue to use DICE as a source of damage formulations, the following adjustments are suggested:

- The quadratic damage formulations for sea level rise and other damages, including their treatment of adaptation, need to be further documented and justified.
- Regional and sectoral damage projection detail needs to be made available either through explicit modeling or a clearly documented calibration.

- The calibration of the individual noncatastrophic impact categories (e.g., agriculture, energy demand, coastal infrastructure, human health) need to be reevaluated in light of recent literature: for documentation, see Nordhaus and Boyer (2000) and Nordhaus (2007); also see discussion in Chapter 5, "Current Literature on Climate Damages."
- Additional types of damages could be considered for inclusion (see Table 5-3 and the discussion in Chapter 5, "Updating Individual Sectoral Damage Functions").
- If the calibration allows for meaningful characterization of parametric uncertainty, the damage functions need to be updated to include parametric uncertainty.
- The catastrophic damages calibration needs to be revisited and updated, if possible, and also revised to represent the stochastic nature of the "catastrophic" damages term.
- The IWG needs to avoid using a damage formulation whose calibration is based on meta-analysis of damage estimates from other SC-IAMs unless it is used by itself or the social cost of carbon estimation approach accounts for this between-model dependence.

FUND

If FUND continues as a source of damage formulations, the committee suggests the following adjustments:

- Further justification is needed for the damage formulations for agriculture, heating demand, cooling demand, and mortality, the assumptions underlying adaptation in the different sectors, the regional distribution of damages, and the parametric uncertainties overall.
- The calibration of the individual noncatastrophic impact categories in FUND (agriculture, energy demand, coastal infrastructure, human health), and their parametric uncertainty, need to be evaluated in light of recent literature and updated, as possible.
- Additional types of damages could be considered for inclusion (see Table 5-3 and the section "Updating Individual Sectoral Damage Functions" in Chapter 5).

PAGE

If PAGE is maintained as a source of damage formulations, the committee suggests the following adjustments:

- PAGE is the least well documented of the three SC-IAMs. Although the structure of the model is described in a number of publications and working papers, the committee was unable to find documentation providing scientific rationales for the parameter distributions used in the damage function. In addition, the code for the model is not publicly available. If the IWG wishes to continue with PAGE as one of the damage formulations going forward, clear documentation needs to be developed and the code needs to be made publicly available.
- The damage formulations, parametric uncertainties, and observed model behavior need further documentation and scientific justification. Particular focus needs to be given to noneconomic, economic, and discontinuity damages, regional distribution and scaling of damages, adaptation modeling and costs, and parametric uncertainties.
- The calibration of PAGE09 is based on damage estimates from other SC-IAMs. The IWG needs to avoid using a damage formulation whose calibration is based on damage estimates from other SC-IAMs unless it is used by itself or the SCC estimation approach accounts for this between-model dependence.

REFERENCES

Nordhaus, W.D., and Boyer, J. (2000). *Warming the World: Economic Models of Global Warming.* Cambridge, MA: MIT Press.
Nordhaus, W.D. (2007). *Accompanying Notes and Documentation on Development of DICE-2007 Model: Notes on DICE-2007.delta.v8 as of September 21, 2007.* New Haven, CT: Yale University Press. Available: http://www.econ.yale.edu/~nordhaus/homepage/Accom_Notes_100507.pdf [October 2016].